Farming and the Fate of Wild Nature

Essays in Conservation-Based Agriculture

Edited by
Daniel Imhoff and Jo Ann Baumgartner

Healdsburg, California

Published by Watershed Media
451 Hudson Street
Healdsburg, California 95448
www.watershedmedia.org

Produced by the Wild Farm Alliance
P.O. Box 2570
Watsonville, California 95077
www.wildfarmalliance.org

Distributed by
University of California Press
Berkeley and Los Angeles, California
University of California Press, Ltd.
London, England
www.ucpress.edu

Cover Photo Credits
Farm scene: Raymond Gehman, National Geographic Image Collection
Bobcat: Susan Morse
Pacific treefrog: Kelly McAllister, Washington Department of Fish and Wildlife
Red-tailed bumblebee: David Cappaet, www.forestryimages.org
Barred owl: John Triana, Regional Water Authority, www.forestryimages.org

Library of Congress Cataloging-in-Publication Data available upon request from the publisher.

Printed in Canada on acid-free, 100 percent post-consumer recycled, processed chlorine-free paper.

ISBN 0-9709500-3-9
First Edition

08 07 06 05
10 9 8 7 6 5 4 3 2 1

Wendell Berry, "Conservationist and Agrarian," in *Citizenship Papers*, Shoemaker & Hoard, 2004.

Richard Manning, "The Oil We Eat: Following the Food Chain Back to Iraq," *Harper's Magazine*, February 2004.

Barbara Kingsolver, "A Forest's Last Stand," in *Small Wonder: Essays*. HarperCollins, 2002.

Ted Williams, "Salmon Stakes." *Audubon*, March 2003.

Gary Paul Nabhan and Ana Valenzuela-Zapata, "When the Epidemic Hit the King of Clones," in *Tequila: A Natural and Cultural History*, The University of Arizona Press, 2004.

Scott McMillon, "Wild Work Crew: Rocky Creek Landowner Puts Resident Beavers to Work," *Bozeman Daily Chronicle*, June 16, 2002.

Luba Vangelova, "Living with Wolves," *Wildlife Conservation Magazine*, March/April 2003.

Daniel Imhoff, "A Plea for Bees," *Vegetarian Times*, June 2005.

Aldo Leopold, "A Biotic View of Land," in *The River of the Mother of God and Other Essays*, The University of Wisconsin Press, 1991.

Dave Foreman, adapted from a forthcoming book, *The Myth(s) of the Environmental Movement: Why Nature Lovers Must Take Back the Conservation Movement.*

Reed Noss, "Context Matters: Considerations for Large-Scale Conservation," *Conservation in Practice*, Summer 2002.

Rick Bass, "Keeping Track," *Forest Magazine*, September/October 2000.

John Terborgh, James Estes, Paul Paquet, Katherine Ralls, Diane Boyd-Heger, Brian Miller, and Reed Noss, "The Role of Top Carnivores in Regulating Terrestrial Ecosystems," in *Continental Conservation*. Island Press, 1991.

Michael Pollan, "The Way We Live Now: The (Agri)Cultural Contradictions of Obesity," *New York Times Magazine*, October 12, 2003.

Brian Halweil, "Can Organic Farming Feed Us All?" *WorldWatch*, May/June 2006.

Dan Barber, "Food Without Fear," *New York Times*, November 23, 2004.

Gretel H. Schueller, David Seideman, Jennifer Bogo, and Angie Jabine. "Taste for Conservation," *Audubon*, May 2004.

CONTENTS

ACKNOWLEDGMENTS

For years, *Farming and the Fate of Wild Nature* remained one of those elusive projects that seemed so obvious and necessary yet so difficult to christen with a final production deadline. But in the publishing world, nothing really seems to count unless it's finished—whatever that truly means—and in early 2006 the Wild Farm Alliance and Watershed Media made this reader a collaborative priority. It is with both excitement and gratitude to so many individuals and organizations that we send this book out to the reading public. Our sincere thanks go to our graphic design team: Timothy Rice for the layout of the book and Kevin Cross for its cover. We were fortunate once again to have Janet Reed Blake as our editor. Christen and Ryan Crumley provided outstanding editorial assistance, and Ellen Sherron produced our index.

This book could not have been possible without the express support from the Foundation for Sustainability and Innovation, Gaia Fund, Garfield Foundation, Patagonia Inc., True North Foundation, and Wallace Genetic Foundation who have so generously stood behind the mission of the Wild Farm Alliance. The great board members of the Wild Farm Alliance, who have remained energized and united since the organization's beginnings in January 2000, need to be heartily acknowledged, and we hope that this volume brings a sense of pride to their ongoing commitment. Thanks also go to Watershed Media's distribution partner, the University of California Press, in particular Hannah Love.

We further wish to thank Merrilee Buchanan, Sam Earnshaw, Quincey Tompkins Imhoff, and Jack Shoemaker for their support behind the scenes; and the board of directors of Watershed Media—Roberto Carra, Jib Ellison, John Harvey, Greg Hendrickson, Diana Donlon Karlenzig, and Anne Telford.

Most of all, we need to thank the many authors, illustrators, and photographers involved, so many of whom donated their work and their skills to give this book its shape and scope. We are truly humbled and grateful and hopeful for a future alive with wildness and the agrarian spirit.

—Daniel Imhoff and Jo Ann Baumgartner, August 2006

INTRODUCTION

A good farm must be one where the native flora and fauna have lost acreage without losing their existence.

—Aldo Leopold

We have collected, mulled over, and carefully considered the following essays over a three-year period and chosen them to provide the scientific, philosophical, economic, and cultural underpinnings for an emerging movement, conservation-based agriculture. A number of the essays also influenced a previous book, *Farming with the Wild,* a project that inspired a continental journey in pursuit of two questions. How much wildnesss can a farm or ranching operation support and still remain economically viable? And how much agriculture can take place in an area and still support optimal levels of biodiversity? The Wild Farm Alliance, the co-producer of this and the earlier book, has been engaged in these questions for nearly a decade and continues to be a leading voice for the re-integration of wildness in farming and ranching regions.

Our world is rapidly changing, and the actions we take today will have far-flung consequences. Many alarming trends cumulatively point to inevitable shifts in agriculture in the years ahead: Changing climate patterns. Limited cheap fossil fuels. Rapid urbanization and rising rates of industrial consumption. Approaching shortages of clean water. The collapse of critical ecosystems overburdened by industrial wastes. The irreparable loss of species, and with them, the planet's evolutionary legacy as well as the genetic resilience to overcome pest outbreaks and other hardships. The pages ahead will reveal a broad platform of conservation-based solutions to many of these challenges.

Industrial agriculture has played an important, if not a leading, role in many of the problems listed above. Forty years ago, the pioneers of our contemporary organic farming movement set out to change that course with a vision for "sustainable agriculture." Throughout four decades of hard work, organic farmers have become extremely successful at growing crops and raising livestock without the toxic chemicals deemed necessary for today's industrial factory-based agriculture. In some areas of the country, many organic farmers and advocates have emerged as leaders in the conservation-based agriculture movement. But as a whole, the organic movement has fallen short of deep reforms because it failed to adequately address the interconnections between farms, ranches, and the landscapes that support them.

One hears, for example, a great deal about "biodiversity" in conversations about sustainable agriculture. This can refer positively to the protection of soil organisms, such as earthworms or mycorrhyzal fungi, both beneficial to farmers. Or it can refer negatively to the devastating loss of traditional crop diversity, in terms of the dwindling numbers, varieties, and breeds of plant and animal species grown and collected for human uses. It is far less often, however, that we hear about "wild biodiversity" in dialogues about sustainable agriculture. By this, we mean the healthy habitats needed to support a wide range of native flora and fauna where agriculture takes place.

This is understandable. Agriculture, after all, involves the domestication of the wild. Over the past three centuries, native habitats—from river valleys and grasslands, to wetlands, uplands, and woodlands—have been converted to agricultural lands. In the industrial economy, agricultural operations have reduced complex landscapes into zones of intensive production for just a handful of exotic crops or, more often, a single monoculture encompassing thousands of acres or feedlot "gulags" housing tens of thousands of animals. In order to compete in global markets, to pay for expensive machinery and inputs, to overcompensate for rising production costs and declining crop prices, or simply to create "clean farms" void of weeds, ever-larger swaths of habitats have been erased. With the clearing of habitat comes the loss of species. And wild biodiversity is pushed farther and farther into isolated pockets on the landscape.

This expansion—primarily to support the grain-fed confinement livestock industry—has sent shock waves across the landscape. As much as two-thirds of all public, private, and tribal lands are now used for agriculture, either in grazing, haying, or row cropping. Half of the wetlands in the continental United States have been lost in the past century. Human activities consume an estimated 40 percent of the earth's daily photosynthetic output, while agriculture uses two-thirds of the earth's available fresh water supplies. According to Defenders of Wildlife, as of 1995, 84 percent of all threatened and endangered plant and animal species in the United States were listed at least partially as a result of agricultural activities. The situation is not that different in other regions of the world.

It is also understandable that the conservation community—even the sustainable agriculture community—has sometimes kept its distance from mainstream agriculture. Yet, ironically perhaps, conservationists, restorationists, and all other citizens concerned about wild biodiversity have little choice but to engage agriculture for solutions to common concerns. The fragmented landscape requires that our existing wild areas (no more than 10 percent of the land base) somehow be reconnected through healthy watersheds and diverse habitat linkages. (A juvenile male mountain lion must safely cross hundreds of miles of territory to become an independent adult; a salmon must find the watershed of its birth to spawn and complete its noble cycle of life.) Many of these linkages and migrations must take place on privately held farms and ranches. Unfortunately, landowners cannot always easily receive compensation or technical assistance for such essential stewardship services.

Advocates of a truly sustainable agriculture must also embrace wild biodiversity as an essential foundation for a new farming and ranching regime, one that works with wild nature rather than industrializing it. Indeed, as shown in many of the essays that follow, some people are already working to combine agriculture and conservation in extraordinary ways.

While the authors of this book may not necessarily agree on approaches and outlook, or on how to achieve a balance between agriculture and conservation goals, it is safe to assume that they share a common value. Healthy human activities require healthy landscapes, and healthy landscapes require moving away from an eradication ethic toward coexistence with all species. The fate of wild nature will no doubt be deeply intertwined with the food and farming systems of our present and future. But this is a codependent journey. There will be no agriculture within completely degraded habitats.

These essays show us that we still know so little about the potential of working with wild nature in highly productive ways. But this is not a call to convert the earth into a "working landscape." Wilderness must continue to be protected for its own sake, as well as to guide and inspire us. The broad concepts of conservation-based agriculture are ours for the taking: A grass farming revolution to replace the corn-soybean-feedlot juggernaut. A continental effort to restore habitat for native pollinators, wide-ranging predators, and vital watersheds. The preservation of migratory salmon populations. Vital farming and ranching regions in which native species may lose acreage without losing existence. These are ideas that just might save us.

—Dan Imhoff and Jo Ann Baumgartner

Section I

AGRICULTURE & CONSERVATION

For some time, many authors have persuasively argued for a more seamless integration of wildness with landscapes domesticated and developed for human economies. Understanding and implementing such an integration is a humbling and revelatory process, a bit like the peeling of an onion. The barriers we have built between the domesticated and the wild, between industrial and biological worlds, are complex and formidable and not so easily undone.

Yet food and farming remain common necessities that can ultimately unite us in that effort. We all need to eat, and our ability to continue to nourish ourselves will require sustained productivity as well as thoughtful stewardship and conservation. To that end, our conservation goals are no doubt better served by a thorough understanding of the economic forces that affect agricultural conditions today. At the same time, our understanding of sustainable agriculture must be oriented around the concept of vital ecological neighborhoods that encompass entire farming regions and include healthy soils, clean watersheds, perennial grasslands, and abundant wildness.

CONSERVATIONIST AND AGRARIAN

WENDELL BERRY

I am a conservationist and a farmer, a wilderness advocate and an agrarian. I am in favor of the world's wildness, not only because I like it, but also because I think it is necessary to the world's life and to our own. For the same reason, I want to preserve the natural health and integrity of the world's economic landscapes, which is to say that I want the world's farmers, ranchers, and foresters to live in stable, locally adapted, resource-preserving communities, and I want them to thrive.

One thing that means is that I have spent my life on two losing sides. As long as I have been conscious, the great causes of agrarianism and conservation, despite local victories, have suffered an accumulation of losses, some of them probably irreparable—while the third side, that of the land-exploiting corporations, has appeared to grow ever richer. I say "appeared" because I think their wealth is illusory. Their capitalism is based, finally, not on the resources of nature, which it is recklessly destroying, but on fantasy. Not long ago I heard an economist say, "If the consumer ever stops living beyond his means, we'll have a recession." And so the two sides of nature and the rural communities are being defeated by a third side that will eventually be found to have defeated itself.

Perhaps in order to survive its inherent absurdity, the third side is asserting its power as never before: by its control of politics, of public education, and of the news media; by its dominance of science; and by biotechnology, which it is commercializing with unprecedented haste and aggression in order to control totally the world's land-using economies and its food supply. This massive ascendancy of corporate power over democratic process is probably the most ominous development since the end of World War II, and for the most part "the free world" seems to be regarding it as merely normal.

My sorrow in having been for so long on two losing sides has been compounded by knowing that those two sides have been in conflict, not only with their common enemy, the third side, but also, and by now almost conventionally, with each other. And I am further aggrieved in understanding that everybody on my two sides is deeply implicated in the sins and in the fate of the self-destructive third side.

As a part of my own effort to think better, I decided not long ago that I would not endorse any more wilderness preservation projects that do not seek also to improve the health of the surrounding economic landscapes and human communities. One of my reasons is that I don't think we can preserve either wildness or wilderness areas if we can't preserve the economic landscapes and the people who use them. This has put me into discomfort with some of my conservation friends, but that discomfort only balances the discomfort I feel when farmers or ranchers identify me as an "environmentalist," both because I dislike the term and because I sympathize with farmers and ranchers.

Whatever its difficulties, my decision to cooperate no longer in the separation of the wild and the domestic has helped me to see more clearly the compatibility and even the coherence of my two allegiances. The dualism of domestic and wild is, after all, mostly false, and it is misleading. It has obscured for us the domesticity of the wild creatures. More important, it has obscured the absolute dependence of human domesticity upon the wildness that supports it and in fact permeates it. In suffering the now-common accusation that humans are "anthropocentric" (ugly word), we forget that the wild sheep and the wild wolves are respectively ovicentric and lupocentric. The world, we may say, is wild, and all the creatures are homemakers within it, practicing domesticity: mating, raising young, seeking food and comfort. Likewise, though the wild sheep and the farm-bred sheep are in some ways unlike in their domesticities, we forget too easily that if the "domestic" sheep becomes too unwild, as some occasionally do, they become uneconomic and useless: They have reproductive problems, conformation problems, and so on. Domesticity and wildness are in fact intimately connected. What is utterly alien to both is corporate industrialism—a dislocated economic life that is without affection for the places where it is lived and without respect for the materials it uses.

The question we must deal with is not whether the domestic and the wild are separate or can be separated; it is how, in the human economy, their indissoluble and necessary connection can be properly maintained. But to say that wildness and domesticity are not separate, and that we humans are to a large extent responsible for the proper maintenance of their relationship, is to come under a heavy responsibility to be practical. I have two thoroughly practical questions on my mind.

The first is: Why should conservationists have a positive interest in, for example, farming? There are lots of reasons, but the plainest is: Conservationists eat. To be interested in food but not in food production is clearly absurd. Urban conservationists may feel entitled to be unconcerned about food production because they are not farmers. But they can't be let off so easily, for they are all farming by proxy. They can eat only if land is farmed on their behalf by somebody somewhere in some fashion. If conservationists will attempt to resume responsibility for their need to eat, they will be led back fairly directly to all their previous concerns for the welfare of nature.

Do conservationists, then, wish to eat well or poorly? Would they like their food supply to be secure from one year to the next? Would they like their food to be free of poisons, antibiotics, alien genes, and other contaminants? Would they like a significant portion of it to be fresh? Would they like it to come to them at the lowest possible ecological cost? The answers, if responsibly given, will influence production, will influence land use, will determine the configuration and the health of landscapes.

If conservationists merely eat whatever the supermarket provides and the government allows, they are giving economic support to all-out industrial food production: to the animal factories; to the depletion of soil, rivers and aquifers; to crop monocultures and the consequent losses of biological and genetic diversity; to the pollution, toxicity, and overmedication that are the inevitable accompaniments of all-out industrial food production; to a food system based on long-distance transportation and the consequent waste of petroleum and the spread of pests and diseases; and to the division of the countryside into ever-larger farms and ever-larger fields receiving always less human affection and human care.

If, on the other hand, conservationists are willing to insist on having the best food, produced in the best way, as close to their homes as possible, and if they are willing to learn to judge the quality of food and food production, then they are going to give economic support to an entirely different kind of land use in an entirely different landscape. This landscape will have a higher ratio of caretakers to acres, of care to use. It will be at once more domestic and more wild than the industrial landscape. Can increasing the number of farms and farmers in an agricultural landscape enhance the quality of that landscape as wildlife habitat? Can it increase what we might call the wilderness value of that landscape? It *can* do so, and the determining factor would be diversity. Don't forget we are talking about a landscape that is changing in response to an increase in local consumer demand for local food. Imagine a modern agricultural landscape devoted mainly to corn and soybeans and to animal factories. And then imagine its neighboring city developing a demand for good, locally grown food. To meet that demand, local farming would have to diversify.

If that demand is serious, if it is taken seriously, if it comes from informed and permanently committed consumers, if it promises the necessary economic support, then that radically oversimplified landscape will change. The crop monocultures and animal factories will give way to the mixed farming of plants and animals. Pastured flocks and herds of meat animals, dairy herds, and poultry flocks will return, requiring, of course, pastures and hayfields. If the urban consumers would extend their competent concern for the farming economy to include the forest economy and its diversity of products, that would improve the quality and care, and increase the acreage, of farm woodlands. And we should not forget the possibility that good farmers might, for their own instruction and pleasure, preserve patches of woodland unused. As the meadows and woodlands flourished in the landscape, so would the wild birds and animals. The acreages devoted to corn and soybeans, grown principally as livestock feed or as raw materials for industry, would diminish in favor of the fruits and vegetables required by human dinner tables.

As the acreage under perennial cover increased, soil erosion would decrease and the water-holding capacity of the soil would increase. Creeks and rivers would grow cleaner and their flow more constant. As farms diversified, they would tend to become smaller because complexity and work increase with diversity, and so the landscape would acquire more owners. As the number of farmers and the diversity of their farms increased, the toxicity of agriculture would decrease—insofar as agricultural chemicals are used to replace labor and to defray the biological costs of monoculture. As food production became decentralized, animal wastes would be dispersed and would be absorbed and retained in the soil as nutrients rather than flowing away as waste and as pollutants. The details of such a transformation could be elaborated almost endlessly. To make short work of it here, we could just say that a dangerously oversimplified landscape would become healthfully complex, both economically and ecologically.

Moreover, since we are talking about a city that would be living in large measure from its local fields and forests, we are talking also about a local economy of decentralized, small, nonpolluting, value-adding factories and shops that would be scaled to fit into the landscape with the least ecological or social disruption. And thus we can also credit to this economy an increase in independent small businesses, in self-employment, and a decrease in the combustible fuel needed for transportation and (I believe) for production.

Such an economy is technically possible, there can be no doubt of that; we have the necessary methods and equipment. The capacity of nature to accommodate, and even to cooperate in, such an economy is also undoubtable; we have the necessary historical examples. This is not, from nature's point of view, a pipe dream.

What *is* doubtable, or at least unproven, is the capacity of modern humans to choose, make, and maintain such an economy. For at least half a century we have taken for granted that the methods of farming could safely be determined by the mechanisms of industry, and that the economies of farming could safely be determined by the economic interests of industrial corporations. We are now running rapidly to the end of the possibility of that assumption. The social, ecological, and even the economic costs have become too great, and the costs are still increasing, all over the world.

Now we must try to envision an agriculture founded, not on mechanical principles, but on the principles of biology and ecology. Sir Albert Howard and Wes Jackson have argued at length for such a change of standards. If you want to farm sustainably, they have told us, then you have got to make your farming conform to the natural laws that govern the local ecosystem. You have got to farm with both plants and animals in as great a diversity as possible, you have got to conserve fertility, recycle wastes, keep the ground covered, and so on. Or, as J. Russell Smith put it seventy years ago, you have got to "fit the farming to the farm"—not to the available technology or the market, as important as those considerations are, but to the land. It is necessary, in short, to maintain a proper connection between the domestic and the wild. The paramount standard by which the work is to be judged is the health of the place where the work is done.

But this is not a transformation that we can just drift into, as we drift in and out of fashions, and it is not one that we should wait to be forced into by large-scale ecological breakdown. It won't happen if a lot of people—consumers and producers, city people and country people, conservationists and land users—don't get together deliberately to make it happen.

Those are some of the reasons conservationists should take an interest in farming and make common cause with good farmers. Now I must get on to the second of my practical questions.

Why should farmers be conservationists? Or maybe I had better ask why *are* good farmers conservationists? The farmer lives and works in the meeting place of nature and the human economy, the place where the need for conservation is most obvious and most urgent. Farmers either fit their farming to their farms, conform to the laws of nature, and keep the natural powers and services intact—or they do not. If they do not, then they increase the ecological deficit being charged to the future. (I had better admit that some farmers do increase the ecological deficit, but they are not the farmers I am talking about. I am not asking conservationists to support destructive ways of farming.)

Good farmers, who take seriously their duties as stewards of Creation and of their land's inheritors, contribute to the welfare of society in more ways than society usually acknowledges, or even knows. These farmers produce valuable goods, of course; but they also conserve soil, they conserve water, they conserve wildlife, they conserve open space, they conserve scenery.

All that is what farmers *ought* to do. But since our present society's first standard in all things is profit and it loves to dwell on "economic reality," I can't resist a glance at these good farmers in their economic circumstances, because these farmers will be poorly paid for the goods they produce, and for the services they render to conservation they will not be paid at all. Good farmers today may market products of high quality and perform well all the services I have listed, and *still* be unable to afford health insurance, and *still* find themselves mercilessly caricatured in the public media as rural simpletons, hicks, or rednecks. And then they hear the voices of the "economic realists": "Get big or get out. Sell out and go to town. Adapt or die." We have had fifty years of such realism in agriculture, and the result has been more and more large-scale monocultures and factory farms, with their ever-larger social and ecological—and ultimately economic—costs.

Why do good farmers farm well for poor pay and work as good stewards of nature for no pay, many of them, moreover, having no hope that their farms will be farmed by their children (for the reasons given) or that they will be farmed by anybody?

Well, I was raised by farmers, have farmed myself, and have in turn raised two farmers—which suggests to me that I may know something about farmers, and also that I don't know very much. But over the years I along with a lot of other people have wondered, "Why do they do it?" Why do farmers farm, given their economic adversities on top of the many frustrations and difficulties normal to farming? And always the answer is: "Love. They must do it for love." Farmers farm for the love of farming. They love to watch and nurture the growth of plants. They love to live in the presence of animals. They love to work outdoors. They love the weather, maybe even when it is making them miserable. They love to live where they work and to work where they live. If the scale of their farming is small enough, they like to work in the company of their children and with the help of their children. They love the measure of independence that farm life can still provide. I have an idea that a lot of farmers have gone to a lot of trouble merely to be self-employed, to live at least a part of their lives without a boss.

And so the first thing farmers as conservationists must try to conserve is their love of farming and their love of independence. Of course they can conserve these things only by handing them down, by passing them on to their children, or to *somebody's* children. Perhaps the most urgent task for all of us who want to eat well and to keep eating is to encourage farm-raised children to take up farming. And we must recognize that this only can be done economically. Farm children are not encouraged by watching their parents take their products to market only to have them stolen at prices less than the cost of production.

But farmers obviously are responsible for conserving much more than agrarian skills and attitudes. I have already told why farmers should be, as much as any conservationist, conservers of the wildness of the world—and that is their inescapable dependence on nature. Good farmers, I believe, recognize a difference that is fundamental between what is natural and what is manmade. They know that if you treat a farm as a factory and living creatures as machines, or if you tolerate the idea of "engineering" organisms, then you are on your way to something destructive and, sooner or later, too expensive. To treat creatures as machines is an error with large practical implications.

Good farmers know too that nature can be an economic ally. Natural fertility is cheaper, often in the short run, always in the long run, than purchased fertility. Natural health, inbred and nurtured, is cheaper than pharmaceuticals and chemicals. Solar energy—if you know how to capture and use it: in grass, say, and the bodies of animals—is cheaper than petroleum. The highly industrialized factory farm is entirely dependent on "purchased inputs." The agrarian farm, well integrated into the natural systems that support it, runs to an economically significant extent on resources and supplies that are free.

It is now commonly assumed that when humans took to agriculture they gave up hunting and gathering. But hunting and gathering remained until recently an integral and lively part of my own region's traditional farming life. People hunted for wild game; they fished the ponds and streams; they gathered wild greens in the spring, hickory nuts and walnuts in the fall; they picked wild berries and other fruits; they prospected for wild honey. Some of the most memorable, and least regrettable, nights of my own youth were spent in coon hunting with farmers. There is no denying that these activities contributed to the economy of farm households, but a further fact is that they were pleasures; they were wilderness pleasures, not greatly different from the pleasures pursued by conservationists and wilderness lovers. As I was always aware, my friends the coon hunters were not motivated just by the wish to tree coons and listen to hounds and listen to each other, all of which were sufficiently attractive; they were coon hunters also because they wanted to be afoot in the woods at night. Most of the farmers I have known, and certainly the most interesting ones, have had the capacity to ramble about outdoors for the mere happiness of it, alert to the doings of the creatures, amused by the sight of a fox catching grasshoppers, or by the puzzle of wild tracks in the snow.

As the countryside has been depopulated and the remaining farmers have come under greater stress, these wilderness pleasures have fallen away. But they have not yet been altogether abandoned; they represent something probably essential to the character of the best farming, and they should be remembered and revived.

Those, then, are some reasons why good farmers are conservationists, and why all farmers ought to be.

What I have been trying to do is define a congruity or community of interest between farmers and conservationists who are not farmers. To name the interests that these two groups have in common, and to observe, as I did at the beginning, that they also have common enemies, is to raise a question that is becoming increasingly urgent: Why don't the two groups publicly and forcefully agree on the things they agree on, and make an effort to cooperate? I don't mean to belittle their disagreements, which I acknowledge to be important. Nevertheless, cooperation is now necessary, and it is possible. If Kentucky tobacco farmers can meet with antismoking groups, draw up a set of "core principles" to which they all agree, and then support those principles, something of the sort surely could happen between conservationists and certain land-using enterprises: family farms and ranches, small-scale, locally owned forestry and forest-products industries, and perhaps others. Something of the sort, in fact, is beginning to happen, but so far the efforts are too small and too scattered. The larger organizations on both sides need to take an interest and get involved.

If these two sides, which need to cooperate, have so far been at odds, what is the problem? The problem, I think, is economic. The small land-users, on the one hand, are struggling so hard to survive in an economy controlled by the corporations that they are distracted from their own economy's actual basis in nature. They also have not paid enough attention to the difference between their always threatened local economies and the apparently thriving corporate economy that is exploiting them.

On the other hand, the mostly urban conservationists, who mostly are ignorant of the economic adversities of, say, family-scale farming or ranching, have paid far too little attention to the connection between *their* economic life and the despoliation of nature. They have trouble seeing that the bad farming and forestry practices that they oppose as conservationists are done on their behalf and with their consent implied in the economic proxies they have given as consumers.

These clearly are serious problems. Both of them indicate that the industrial economy is not a true description of economic reality, and moreover that this economy has been wonderfully successful in getting its falsehoods believed. Too many land users and too many conservationists seem to have accepted the doctrine that the availability of goods is determined by the availability of cash, or credit, and by the market. In other words, they have accepted the idea always implicit in the arguments of the land-exploiting corporations: that there can be, and that there is, a safe disconnection between economy and ecology, between human domesticity and the wild world. Industrializing farmers have too readily assumed that the nature of their land could safely be subordinated to the capability of their technology, and that conservation could safely be left to conservationists. Conservationists have too readily assumed that the integrity of the natural world could be preserved mainly by preserving tracts of wilderness, and that the nature and nurture of the economic landscapes could safely be left to agribusiness, the timber industry, debt-ridden farmers and ranchers, and migrant laborers.

To me, it appears that these two sides are as divided as they are because each is clinging to its own version of a common economic error. How can this be corrected? I don't think it can, so long as each of the two sides remains closed up in its own conversation. I think the two sides need to enter into *one* conversation. They have got to talk to one another. Conservationists have got to know and deal competently with the methods and economics of land use. Land users have got to recognize the urgency, even the economic urgency, of the requirements of conservation.

Failing this, these two sides will simply concede an easy victory to their common enemy, the third side, the corporate totalitarianism that is now rapidly consolidating as "the global economy," and that will utterly dominate both the natural world and its human communities.

TAME AND WILD

FREDERICK KIRSCHENMANN AND DAVID GOULD

...the boundary between...tame and wild, exists only in the imperfections of the human mind.

— Aldo Leopold

Organic farms often are erroneously perceived as isolated oases disconnected from surrounding landscapes. This perspective lends itself to several unfortunate misconceptions. It suggests that a farm can be healthy apart from the health of the ecosystem in which it is embedded. It focuses attention on a small segment of the ecosystem instead of the larger watersheds and ecological regions that to a great extent determine the health of an ecosystem. And it leads us to believe that human intention and intervention can segregate and control the farm's productive environment.

"ORGANIC" AS ECOLOGICAL COMMUNITY

This perspective ignores the discoveries of 20th century science that all of nature is interconnected. Everything in nature—plants, animals, soil, water, air—operates in a web of interdependency; a tightly knit network of competition and cooperation. Larry Rasmussen put it succinctly: "The scientific discovery of the 20th century is that the earth is a *community*." Citing Brian Swimme and Thomas Berry, Rasmussen asserts that the earth is "organic" in that the "inner coherence and integral functioning of the Planet" is so unified that "every aspect of the Earth is affected by what happens to any component member of the community." Accordingly, the "well-being of the planet is the condition of the well-being of its member communities. Jeopardy for the one is jeopardy for the other." If Rasmussen is correct, then the use of the word *organic* to designate a small, isolated patch of this dynamic, emerging, interwoven community as somehow separate from the whole may actually be a misuse of the term.

The organic community has, of course, paid tribute to the impossibility of isolating organic farms from the rest of the biotic community. While the act of certifying an organic farm assumes that human intention and intervention can isolate and control a small patch of an ecosystem, certifiers have always stopped short of assuring such isolation. Organic certifiers have never guaranteed consumers residue-free food. They only verify that management practices used on a farm are consistent with organic standards. This is a tacit admission, for example, that "drift" (the movement of materials unacceptable in organic production from a nonorganic field to an organic one) cannot be entirely controlled in an interconnected world. And as every organic farmer knows, the problem of drift has been significantly exacerbated with the introduction of transgenic crops (GMOs). In a complex, dynamic, emergent, biotic community, the inadvertent transfer of unwanted *biological* organisms onto an organic farm is much more potent than the transfer of *chemical* compounds.

While drift (introducing unwanted materials onto an organic farm) is the *negative* aspect of the web of life, the *positive* aspect is the rich biodiversity provided by that same web, which is essential to the biological health of any farm. The biodiversity necessary to produce the ecosystem services on which *all* farms depend can be restored and maintained only through the interaction of the diversity of species at the ecosystem level. It is the co-evolution of a diverse array of species interacting with one another within ecosystems that gives nature its dynamic resilience—a phenomenon Stuart Kauffman calls "interacting dancing fitness landscapes."

A number of years ago a few organic certifiers began to acknowledge this concept. For example, some U.S. certifiers included specific standards to enhance biodiversity so that their certified products can be sold in Switzerland under the Bio-Suisse certification seal. Bio-Suisse recognizes the importance of restoring natural habitat that mimics wildness. It mandates the inclusion of shelterbelts for wild habitat as part of its *production* standards.

More recently in the United States, the Wild Farm Alliance has been instrumental in helping the organic community address biodiversity conservation requirements that were written in the USDA's national organic rule (NOP) but ignored until recently. With their help, the National Organic Standards Board adopted a set of biodiversity inspections questions as part of their model organic system plan. Organic certifiers across the country are now beginning to incorporate on-farm biodiversity enhancement considerations into their inspection processes/criteria/requirements.

While Bio-Suisse was among the first certifiers to require mini-refuges, its persistence on this issue led other certifiers in Europe to adopt similar standards. In fact, the Basic Standards of the International Federation of Organic Agriculture Movements has also committed to this principle. The Standards mandate that organic production systems make a beneficial contribution to the wildness of the ecosystems of which they are a part. "Operators should maintain a significant portion of their farms to facilitate biodiversity and nature conservation.... These include...wildlife corridors that provide linkages and connectivity to native habitat." As of late 2006, IFOAM also has a very well thought out draft set of biodiversity standards that go beyond what is currently required. Only time will tell whether and how they may be implemented.

While standards requiring the restoration of wildness in ecosystems may seem appealing in countries such as Switzerland with relatively small acreages, they are even more important in areas with more extensive agricultural land tracts, such as the United States, Brazil, and Argentina, where field boundaries, shelterbelts, and riparian buffers increasingly give way to specialized, uniform, simplified field management. In the absence of conscious efforts to maintain wild species habitat on such landscapes, human cultivation tends to exclude all species save those we wish to harvest. Eventually the species diversity that maintains the entire ecosystem's capacity for self-renewal disappears. And along with it, we lose the free ecosystem values and benefits that such biodiversity provides to agriculture. Organic farms that fail to pay attention to such landscape issues will fare no better than their nonorganic neighbors. Patches of organic production, regulated by standards that are focused primarily on materials and practices that are not allowed, ignore the importance of the web of interdependence in nature.

As Niles Eldredge puts it:

It is not just a simple question of commandeering the photosynthetic capabilities of a few plant species and forgetting about everything else out there. We still very much that "else"—all those other kinds of organisms still out there in the time-honored, primordial way.

All of this suggests that our task as organic farmers is not simply a matter of commandeering the living organisms we need to produce healthy, organic foods on our farms and forgetting about everything "else." Organic agriculture must begin to reinvent itself to care for landscape ecologies. If we are to be successful as organic farmers in the long term, we cannot ignore the intricate and complex ecological processes of nature that sustain the watersheds, ecological regions, and ecosystems within which our farms exist. Farms can be healthy only when they are part and parcel of healthy "wild" ecological neighborhoods. And as we restore such landscape health, we can begin to harvest "organic" products from them in a sustainable manner.

Aldo Leopold continually reminded us that there are no boundaries between tame and wild, except in the imperfections of our minds. He also pointed out that we cannot reasonably designate one species as valuable, another as harmless, and another as injurious because they are all part of "a biota so complex, so conditioned by interwoven cooperations and competitions, that no man can say where utility begins or ends ...the only sure conclusion is that the biota as a whole is useful...the function of species is largely inscrutable, and may remain so." It is all part of the co-mutual dance of life, and either all the dance is moving toward health or none of it is.

RELATING TO WILDERNESS IN NORTH AMERICA

For those North Americans of European descent, the relationship between tame and wild has always been contentious. When the Puritans first landed on the shores of New England it was with the clear purpose of "taming the wilderness" and establishing "the Kingdom of God on earth." Wildness to their minds was, at best, suspect. In fact, for them, leaving wilderness "unimproved" was a clear indication that one was derelict in one's duty to God. Taking the land from its native dwellers often was justified on the grounds that the Indians had failed to exercise their responsibility to "do something" with the land God had given them. Cotton Mather, in fact, often referred to wilderness as "the Devil's playground," so who could object to taking "wild" land away from the devil and transferring it to "the Kingdom of God"?

Of course the Puritans' principal aim was to create a "Holy Commonwealth" (a moral and just social order). That is what they felt God had specifically chosen them to do. But they also believed in immediate retribution. If they were faithful to their mandate, God would reward them with economic success; if they weren't they would experience economic failure. This pragmatic test meant that establishing a Holy Commonwealth became indistinguishable from clearing the forests and plowing up the prairie in order to grow corn and other domesticated crops for economic benefit. It was all part of their understanding of the special destiny for which they had been chosen.

This Puritan frame of mind was, of course, not limited to the few boatloads of English settlers who came ashore in New England. It has become part of our national ethic. As Sidney Mead reminds us, "this strong sense of particular calling, of destiny under God, has remained a constant part of the ideological structure of the nation." For the original settlers, and to a large extent for us, trees, grass, and wilderness generally "stood in the way, not only of progress, but of deeper notions of order and light."

WILD ORGANIC CERTIFICATION: THE CONTROVERSY

Given this cultural history it is easy to understand why it may be so hard for us to wrap our minds around a unified concept that views tame and wild as part of the same fabric and why our relationship to wildness is still so fraught with controversy.

Late in 1999 some of the private North American certifiers announced that they were working with fisheries in Alaska to determine whether or not wild harvested fish in some of the pristine waters of part of the Bering Sea could qualify for organic certification. This generated an immediate firestorm of debate.

On the one hand, the heated disagreement was perhaps inevitable. The certification of organic products emerged in the industrial world where human ingenuity is believed to be superior to nature. And given all the effort that went into developing organic agriculture and securing its prominence in the marketplace, it was understandable that practitioners would be protective of what they had developed and suspicious of anything that was produced without significant human intention and intervention.

On the other hand, it seems ironic that a controversy would occur over the inclusion of wild production in organic certification since organic production systems have historically been viewed as systems that *mimic nature.* Sir Albert Howard, one of the icons of the organic agriculture movement, went so far as to suggest that nature herself is actually engaged in farming and that farmers should use *nature's* farming methods as the model for their own husbandry. This raises an interesting question for the organic movement. If nature's own wild farming systems are the model that organic farmers emulate, how can we erect a wall of separation between the two systems and assert that the imitation is certifiable but the original is not?

ORGANIC CERTIFICATION AS A MARKET SYSTEM

Still, one can understand the reluctance of organic practitioners to include wild systems in the organic arena. Organic certification has tended to focus on production enclaves isolated from the rest of nature by clear boundaries. Usually "buffer zones" are required to make certain that the enclaves are not contaminated from the outside. It is then assumed that sufficient control can be exercised to ensure that only practices and inputs acceptable by organic standards are used inside those enclosed areas.

This approach has enabled the organic movement to provide reasonable assurance that organic standards are technically followed and that consumers who purchase food produced from such enclosed areas can buy such products with confidence. It is a neat system of isolated patches of "kingdom of God" farming thriving in the midst of a landscape that is judged to be unacceptable. And, of course, given the extent to which unacceptable materials and practices are used on surrounding nonorganic farms, the landscape *is* unacceptable.

There are several problems with this scenario. In the first place it fails to give sufficient recognition to the fact that within nature it is impossible to maintain closed systems. The "patch" ecology approach, inherent in certifying the enclaves of production that we call our "organic" farms, ignores John Muir's observation (now universally accepted) that in nature *everything* is connected. Patch ecology, we now know, simply doesn't work. We will likely learn this incontrovertible lesson once again as we suggest that transgenic crops and nontransgenic crops can "coexist" so long as we put sufficiently stringent restrictions into place. But, in nature the rain, wind, currents, inversions, insects, and other natural organisms, all carry some particles of materials that do not qualify for organic certification onto organic fields from the surrounding landscape. And animals, even when they are confined with sturdy fences, occasionally escape their organic fields and graze on neighboring nonorganic fields. We simply cannot effectively separate one piece of nature from another, or the tame from the wild.

Furthermore, all species in an ecosystem have co-evolved with links of interdependency that we have barely begun to understand. The science of networks has further enriched our thinking in this regard. This new science has called our attention to the fact that "weak links" in a system (those we usually ignore because they appear to play insignificant roles in the resilience of systems) are often the very links that keep a system functioning.

Ultimately there is no way that we can create the rich biodiversity nature needs for healthy functioning behind our walled-in patches of organic production. (Witness the failed $200 million Biosphere II experiment in Arizona a few years ago.) Healthy functioning ecologies need the diversity of species that can only be produced in a sufficiently diverse habitat, and such habitats *require* significant wildness.

The key concepts behind the certification of wild areas are the careful observation, appreciation, and protection of the areas in question and the absence of any human interference other than harvest. Only by recognizing, valuing, and learning to live harmoniously with wildness at the local level can we truly bridge the chasm between human culture and nature. This suggests that the proper role of the human species within the biotic community may be one of appreciation, restoration, and adaptation, rather than intention, intervention, and control.

Our determination to *conquer* nature (implicit in the world-views of both the Puritans and 17th-century science) is misguided. That view was based on a mechanical perception of nature in a static state of equilibrium. We now know that nature is dynamic, and always evolving (sometimes with bursts of rapid change), and that it is self-organizing. And as Aldo Leopold put it, in this new world of nature as biotic community, we are *not* "conquerors of the land community" but "plain members and citizens of it." Humans must stop trying to control nature, rediscover how they fit into its dynamic, evolving biota, and *adapt.*

NEW VISIONS FOR ORGANIC AGRICULTURE

The earth is a community that has evolved over some four billion years. During that time, the "community" has undergone many changes, including major species extinctions. Following each period of extinction it took nearly ten million years for the community to re-evolve into something like the complex, interconnected, interdependent, dynamic, emerging biotic community we are part of today. Since we are now in the midst of a new human-induced species extinction (some of it caused by our agricultural practices), it is imperative that we take steps to restore and maintain our community. "Maintaining" does not mean keeping it the way it is, since nature, despite our efforts, will continue to evolve and change. Rather, it means retaining the lush diversity of species that has evolved into our current rich biotic community—a community that has demonstrated an enormous capacity for self-renewal.

It is that capacity for self-renewal that engenders the health of both our farms and the wildness that supports them. And, as Aldo Leopold reminded us, enhancing this capacity for self-renewal is our primary conservation responsibility. The more we attend to enhancing this capacity for self-renewal the longer the current community of species can survive, and the *longer* they survive the more it assures their *continued* survival. Per Bak wrote, "The longer a species has been in existence, the longer we can expect it to be around in the future. Cockroaches are likely to outlast humans."

Doing agriculture as if it were isolated from that community (as we have done intentionally in industrial agriculture, and less intentionally, but perhaps inadvertently, in organic agriculture) invites collapse, not only of agriculture, but also of the entire community. Agriculture, no matter how inventive, is ultimately dependent on the ecosystem services provided by a healthy, robust, wild ecosystem—and so are we.

Farming in nature's image, as the founders of the organic agriculture movement envisioned is, therefore, the right approach. But what nature teaches us is that we cannot farm in nature's image in isolated patches of nature, it can be done only in concert with the dynamic, evolving diversity of species supported by wildness.

The kind of future that we need to envision for organic agriculture, therefore, is one in which we take note of the ecological regions in our ecosystem, and learn as much as we can about how the ecologies of those regions function and how we can fit agriculture into them by effectively using the ecosystem services they provide. Such services include everything from micro-organisms that restore soil quality, to predators that keep pests in check, to plants that invite predators to control specific pest problems, to habitats that invite native pollinators—and much more.

Expanding our vision of organic agriculture from organic farms to ecological landscapes holds great promise, not only for our ecological neighborhoods, but for the organic industry as well. Instead of limiting the healing potential of ecological farming practices to the boundaries of our farms (or our bodies), we can begin envisioning the restoration of whole landscapes (both tame and wild) and be economically rewarded for doing so.

We can hardly imagine the ecological and economic benefits to be derived from:

- Restoring habitats in our ecological neighborhoods that bring back alkali bees to pollinate alfalfa fields.
- Rethinking soil and water conservation as a watershed, ecological region, and ecosystem—rather than a farm—project.

- Re-establishing nutrient cycling on a watershed scale, instead of an individual farm-scale.
- Sustainably harvesting a variety of elegant, delicious products for local food markets from the restored habitats "between" our farm fields.
- Attending to the complex local ecosystems that support wild fish habitats and harvesting fish from them in ways that retain the rich biodiversity of species on which the fish depend.

And much more.

In other words, farming in nature's image would not just mean trying to organize a farm along the same principles by which nature operates. It would attempt, as my colleagues at the Wild Farm Alliance insist, to fold agriculture into a landscape that mimics nature's own evolutionary processes—and as a result restore wildness to the land and make agriculture more sustainable.

In this scenario, organic systems and wild systems no longer look so different. And that provides us with an opportunity to rethink our relationship to wildness.

MAKING OUR PEACE WITH WILDNESS

All of this suggests that we should begin removing the boundaries we have created between the tame and the wild. Solving the difficult problems of certifying wild areas can help prepare us for this new future in organic agriculture. In that future we would certify ecological landscapes instead of isolated organic farms, and similarly certify wild areas that meet the standards for organic production. The technical distinctions we have created between wild areas and domesticated areas would largely disappear.

This in *no way* suggests that we abandon our ongoing efforts to set aside and preserve wilderness areas that are largely protected from human activity. All of the reasons why we need to do so, articulated over the years by everyone from John Muir to Wendell Berry, still apply. But as Wes Jackson reminds us continually, we can't just preserve wilderness areas and let everything else go to hell. Either it is all holy or none of it is holy. The idea here is not to domesticate wilderness areas, but to recognize the importance of wildness in our domesticated areas and restore at least some of that wildness.

Freed of the rigid dichotomy between tame and wild in our minds, the organic movement could begin by expanding the idea of "mini-refuges" on organic farms. A farmer might decide to harvest ants from the colonies in a shelter belt between two organic fields (a mini-wild area) and sell them in a gourmet shop specializing in specialty foods. Who would want to argue that the ants (dipped in organic chocolate if you like) could not be labeled certified organic? And if the ants can, in principle, be labeled organic, why not do the same for other wild products of animal origin? The issue quickly becomes a question of scale. Thus the problem is moved from the philosophical to the practical, becoming a matter of *certifiability* rather than *standards.* In other words, it is no longer a question of compatibility but of verification.

Suppose, then, that an operator was certified to harvest wild blueberries for the organic market from a landscape no more contaminated by human pollution than an organic farm. And suppose that the operator decided to harvest deer from that same landscape and the certifier could verify that the deer, owing to territorial behavior, spent their entire lives in the confines of that landscape. Why couldn't the venison be targeted for the organic market? Assuming that the entire area was certifiable, and that the hunting was carried out in a manner that was sustainable (i.e., did not disrupt the deer population in relationship to other species on the landscape, and did not contaminate the unspoiled qualities of the area), on what grounds would one deny an organic label?

Of course such certification becomes much more complicated than the certification processes with which we are familiar. It raises many new questions. How does one certify an area or a region? What inspection criteria should be used? What kind of community of people does one involve in the inspection and certification process? How do we determine if violations occur? These and many other questions have to be answered as we determine how to certify the "organic" ecological landscapes of the future.

But certifiers must not only face these questions when certifying wild areas. They increasingly face the same questions in certifying traditional organic farms when those farms are bounded by neighboring farms that employ activities incompatible with organic certification. Fortunately, a few certifiers have already begun to ask and answer those questions in a rigorous manner as they struggle with the certification of wild areas. The results may eventually also serve as guidelines for *all* organic certification.

For example, the certification standards and policies of Farm Verified Organic, an international certification program with over 25 years experience, have developed elaborate requirements for such certification. They call for diligent identification, research, and ongoing monitoring of the target species and wild area. This includes activities undertaken not only by the operator who wishes to be certified, but also by third-party supervisory and/or research organizations. Life cycles, populations, interspecies interactions, human impacts, and general ecosystem dynamics must all be addressed in the certification process, with claims supported by documented research. Potential contamination, harvest levels compatible with species sustainability, and verifiability of all claims are mandated. Since nature is in a constant state of flux, study and monitoring of the ecosystem must be an ongoing process, and certification could be withdrawn at any time that information comes to light indicating noncompliance with any aspect of the standards. A variation on this novel approach to certification was recently addressed in the NOSB's model organic system plan.

As with any organic certification scenario, the rigor of standards needs to be matched by rigorous enforcement. To assume that certification protocols designed for wild areas are easy to come by, given the state of our largely polluted planet, is to undermine the message of this paper. But we can use the cases of legitimately certifiable wild areas to re-imagine our certification tasks, to re-invent our practice of organic agriculture, and to encourage the restoration of wild areas. By using these processes to reward operators in the marketplace for good stewardship, we can begin to build constituencies to restore wild resources and put us on the long road toward landscape ecological health in every local ecosystem. Each local landscape (both wild and tame) can serve as an example of adapting human activity to nature's functioning, resulting in vibrant "interacting dancing fitness landscapes" throughout the planet, a new planet, where we recognize that tame and wild are intimately related.

THE OIL WE EAT

RICHARD MANNING

We learn as children that there is no free lunch, that you don't get something from nothing, that what goes up must come down, and so on. The scientific version of these verities is only slightly more complex. As James Prescott Joule discovered in the nineteenth century, there is only so much energy.

Special as we humans are, we get no exemptions from the rules. All animals eat plants or eat animals that eat plants. This is the food chain, and pulling it is the unique ability of plants to turn sunlight into stored energy in the form of carbohydrates, the basic fuel of all animals.

Scientists have a name for the total amount of plant mass created by Earth in a given year, the total budget for life. They call it the planet's "primary productivity." We humans, a single species among millions, consume about 40 percent of Earth's primary productivity, 40 percent of all there is. This simple number may explain why the current extinction rate is 1,000 times that which existed before human domination of the planet. We six billion have simply stolen the food, the rich among us a lot more than others.

If you follow the energy, eventually you will end up in a field. Humans engage in a dizzying array of artifice and industry. Nonetheless, more than two-thirds of humanity's cut of primary productivity results from agriculture, two-thirds of which in turn consists of three plants: rice, wheat, and corn. In the 10,000 years since humans domesticated these grains, their status has remained undiminished, most likely because they are able to store solar energy in uniquely dense, transportable bundles of carbohydrates. They are to the plant world what a barrel of refined oil is to the hydrocarbon world. Indeed, aside from hydrocarbons they are the most concentrated form of true wealth—sun energy—to be found on the planet.

The maintenance of such a concentration of wealth often requires violent action. Agriculture is a recent human experiment. For most of human history, we lived by gathering or killing a broad variety of nature's offerings. Why humans might have traded this approach for the complexities of agriculture is an interesting and long-debated question, especially because the skeletal evidence clearly indicates that early farmers were more poorly nourished, more disease-ridden and deformed, than their hunter-gatherer contemporaries. Farming did not improve most lives. The evidence that best points to the answer, I think, lies in the difference between early agricultural villages and their pre-agricultural counterparts—the presence not just of grain but of granaries and, more tellingly, of just a few houses significantly larger and more ornate than all the others attached to those granaries. Agriculture was not so much about food as it was about the accumulation of wealth. It benefited some humans, and those people have been in charge ever since.

Domestication was also a radical change in the distribution of wealth within the plant world. Plants can spend their solar income in several ways. The dominant and prudent strategy is to allocate most of it to building roots, stem, bark—a conservative portfolio of investments that allows the plant to better gather energy and survive the downturn years. Further, by living in diverse stands (a given chunk of native prairie contains maybe 200 species of plants), these perennials provide services for one another, such as retaining water, protecting one another from wind, and fixing free nitrogen from the air to use as fertilizer. Diversity allows a system to "sponsor its own fertility," to use visionary agronomist Wes Jackson's phrase. This is the plant world's norm.

There is a very narrow group of annuals, however, that grow in patches of a single species and store almost all of their income as seed, a tight bundle of carbohydrates easily exploited by seed eaters such as ourselves. Under normal circumstances, this eggs-in-one-basket strategy is a dumb idea for a plant. But not during catastrophes such as floods, fires, and volcanic eruptions. Such catastrophes strip established plant communities and create opportunities for wind-scattered entrepreneurial seed bearers. It is no accident that no matter where agriculture sprouted on the globe, it always happened near rivers. You might assume, as many have, that this is because the plants needed the water or nutrients. Mostly this is not true. They needed the power of flooding, which scoured landscapes and stripped out competitors. Nor is it an accident, I think, that agriculture arose independently and simultaneously around the globe just as the last ice age ended, a time of enormous upheaval when glacial melt let loose sea-size lakes to create tidal waves of erosion. It was a time of catastrophe.

Corn, rice, and wheat are especially adapted to catastrophe. In the natural scheme of things, a catastrophe would create a blank slate, bare soil, that was good for them. Then, under normal circumstances, succession would quickly close that niche. The annuals would colonize. Their roots would stabilize the soil, accumulate organic matter, provide cover. Eventually the catastrophic niche would close. Farming is the process of ripping that niche open again and again. It is an annual artificial catastrophe, and it requires the equivalent of three or four tons of TNT per acre for a modern American farm. Iowa's fields require the energy of 4,000 Nagasaki bombs every year.

When we say the soil is rich, it is not a metaphor. It is as rich in energy as an oil well. A prairie converts that energy to flowers and roots and stems, which in turn pass back into the ground as dead organic matter. The layers of topsoil build up into a rich repository of energy, a bank. A farm field appropriates that energy, puts it into seeds we can eat. Much of the energy moves from the earth to the rings of fat around our necks and waists. And much of the energy is simply wasted, a trail of dollars billowing from a burglar's satchel.

I've already mentioned that we humans take 40 percent of the globe's primary productivity every year. You might have assumed we and our livestock eat our way through that volume, but this is not the case. Part of that total—almost a third of it—is the potential plant mass lost when forests are cleared for farming or when tropical rain forests are cut for grazing or when plows destroy the deep mat of prairie roots that held the whole business together, triggering erosion. The Dust Bowl was no accident of nature. A functioning grassland prairie produces more biomass each year than does even the most technologically advanced wheat field. The problem is, it's mostly a form of grass and grass roots that humans can't eat. So we replace the prairie with our own preferred grass, wheat. Never mind that we feed most of our grain to livestock, and that livestock is perfectly content to eat native grass. And never mind that there likely were more bison produced naturally on the Great Plains before farming than all of beef farming raises in the same area today. Our ancestors found it preferable to pluck the energy from the ground and when it ran out move on.

Today we do the same, only now when the vault is empty we fill it again with new energy in the form of oil-rich fertilizers. Oil is annual primary productivity stored as hydrocarbons, a trust fund of sorts, built up over many thousands of years. On average, it takes 5.5 gallons of fossil energy to restore a year's worth of lost fertility to an acre of eroded land—in 1997 we burned through more than 400 years' worth of ancient fossilized productivity, most of it from someplace else. Even as the earth beneath Iowa shrinks, it is being globalized.

Six thousand years before sodbusters broke up Iowa, their Caucasian blood ancestors broke up the Hungarian plain, an area just northwest of the Caucasus Mountains. Archaeologists call this tribe the LBK, short for *linearbandkeramik,* the German word that describes the distinctive pottery remnants that mark their occupation of Europe. Anthropologists call them the wheat-beef people, a name that better connects those ancients along the Danube to my fellow Montanans on the Upper Missouri River. These proto-Europeans had a full set of domesticated plants and animals, but wheat and beef dominated. All the domesticates came from an area along what is now the Iraq-Syria-Turkey border at the edges of the Zagros Mountains. This is the center of domestication for the Western world's main crops and livestock, ground zero of catastrophic agriculture.

Two other types of catastrophic agriculture evolved at roughly the same time, one centered on rice in what is now China and India and one centered on corn and potatoes in Central and South America. Rice, though, is tropical and its expansion depends on water, so it developed only in floodplains, estuaries, and swamps. Corn agriculture was every bit as voracious as wheat; the Aztecs could be as brutal and imperialistic as Romans or Brits, but the corn cultures collapsed with the onslaught of Spanish conquest. Corn itself simply joined the wheat-beef people's coalition. Wheat was the empire builder; its bare botanical facts dictated the motion and violence that we know as imperialism.

The wheat-beef people swept across the western European plains in less than 300 years, a conquest some archaeologists refer to as a "blitzkrieg." A different race of humans, the Cro-Magnons—hunter-gatherers, not farmers—lived on those plains at the time. Their cave art at places such as Lascaux testifies to their sophistication and profound connection to wildlife. They probably did most of their hunting and gathering in uplands and river bottoms, places the wheat farmers didn't need, suggesting the possibility of coexistence. That's not what happened, however. Both genetic and linguistic evidence say that the farmers killed the hunters. The Basque people are probably the lone remnant descendants of Cro-Magnons, the only trace.

Wheat is temperate and prefers plowed-up grasslands. The globe has a limited stock of temperate grasslands, just as it has a limited stock of all other biomes. On average, about 10 percent of all other biomes remain in something like their native state today. Only one percent of temperate grasslands remains undestroyed.

The supply of temperate grasslands lies in what are today the United States, Canada, the South American pampas, New Zealand, Australia, South Africa, Europe, and the Asiatic extension of the European plain into the sub-Siberian steppes. This area largely describes the First World, the developed world. Temperate grasslands make up not only the habitat of wheat and beef but also the globe's islands of Caucasians, of European surnames and languages. In 2000 the countries of the temperate grasslands, the neo-Europes, accounted for about 80 percent of all wheat exports in the world, and about 86 percent of all corn. That is to say, the neo-Europes drive the world's agriculture. The dominance does not stop with grain. These countries, plus the mothership—Europe—accounted for three-fourths of all agricultural exports of all crops in the world in 1999.

Plato wrote of his country's farmlands:

What now remains of the formerly rich land is like the skeleton of a sick man. . . .

Plato's lament is rooted in wheat agriculture, which depleted his country's soil and subsequently caused the series of declines that pushed centers of civilization to Rome, Turkey, and western Europe. By the fifth century, though, wheat's strategy of depleting and moving on ran up against the Atlantic Ocean. Fenced-in wheat agriculture is like rice agriculture. It balances its equations with famine. In the millennium between 500 and 1500, Britain suffered a major "corrective" famine about every ten years; there were seventy-five in France during the same period. The incidence, however, dropped sharply when colonization of the Americas brought an influx of new food to Europe. The precolonial famines of Europe raised the question: What would happen when the planet's supply of arable land ran out? We have a clear answer. In about 1960 expansion hit its limits and the supply of unfarmed, arable lands came to an end. There was nothing left to plow. What happened was grain yields tripled.

The accepted term for this strange turn of events is the green revolution, though it would be more properly labeled the amber revolution, because it applied exclusively to grain—wheat, rice, and corn. Plant breeders tinkered with the architecture of these three grains so that they could be hypercharged with irrigation water and chemical fertilizers, especially nitrogen. This innovation meshed nicely with the increased "efficiency" of the industrialized factory-farm system. With the possible exception of the domestication of wheat, the green revolution is the worst thing that has ever happened to the planet.

For openers, it disrupted long-standing patterns of rural life worldwide, moving a lot of no-longer-needed people off the land and into the world's most severe poverty. The experience in population control in the developing world is by now clear: It is not that people make more people so much as it is that they make more poor people. In the forty-year period beginning about 1960, the world's population doubled, adding virtually the entire increase of 3 billion to the world's poorest classes, the most fecund classes. The way in which the green revolution raised that grain contributed hugely to the population boom, and it is the weight of the population that leaves humanity in its present untenable position.

Discussion of these, the most poor, however, is largely irrelevant to the American situation. We say we have poor people here, but almost no one in this country lives on less than one dollar a day, the global benchmark for poverty. It marks off a class of about 1.3 billion people, the hard core of the larger group of 2 billion chronically malnourished people—that is, one third of humanity. We may forget about them, as most Americans do.

More relevant here are the methods of the green revolution, which added orders of magnitude to the devastation. By mining the iron for tractors, drilling the new oil to fuel them and to make nitrogen fertilizers, and by taking the water that rain and rivers had meant for other lands, farming had extended its boundaries, its dominion, to lands that were not farmable. At the same time, it extended its boundaries across time, tapping fossil energy, stripping past assets.

The common assumption these days is that we muster our weapons to secure oil, not food. There's a little joke in this. Ever since we ran out of arable land, food is oil. Every single calorie we eat is backed by at least a calorie of oil, more like ten. In 1940 the average farm in the United States produced 2.3 calories of food energy for every calorie of fossil energy it used. By 1974 (the last year in which anyone looked closely at this issue), that ratio was 1:1. And this understates the problem, because at the same time that there is more oil in our food there is less oil in our oil. A couple of generations ago we spent a lot less energy drilling, pumping, and distributing than we do now. In the 1940s we got about 100 barrels of oil back for every barrel of oil we spent getting it. Today each barrel invested in the process returns only ten, a calculation that no doubt fails to include the fuel burned by the Hummers and Blackhawks we use to maintain access to the oil in Iraq.

David Pimentel, an expert on food and energy at Cornell University, has estimated that if all of the world ate the way the United States eats, humanity would exhaust all known global fossil-fuel reserves in just over seven years. Pimentel has his detractors. Some have accused him of being off on other calculations by as much as 30 percent. Fine. Make it ten years.

Agriculture's increasing dependence on petroleum-derived chemicals—polychlorinated biphenyls, polyvinyls, DDT, 2-4d, and other polysyllabic organic compounds—is well known if under-regulated. An infant born in a rural, wheat-producing county in the United States has about twice the chance of suffering birth defects as one born in a rural place that doesn't produce wheat, an effect researchers blame on chlorophenoxy herbicides. Focusing on pesticide pollution, though, misses the worst of the pollutants. It is nitrogen—the wellspring of fertility relied on by every Eden-obsessed backyard gardener and suburban groundskeeper—that we should fear most.

Nitrogen can be released from its "fixed" state as a solid in the soil by natural processes that allow it to circulate freely in the atmosphere. This also can be done artificially. Indeed, humans now contribute more nitrogen to the nitrogen cycle than the planet itself does. That is, humans have doubled the amount of nitrogen in play.

This has led to an imbalance. It is easier to create nitrogen fertilizer than it is to apply it evenly to fields. When farmers dump nitrogen on a crop, much is wasted. It runs into the water and soil, where it either reacts chemically with its surroundings to form new compounds or flows off to fertilize something else, somewhere else.

That chemical reaction, called acidification, is noxious and contributes significantly to acid rain. One of the compounds produced by acidification is nitrous oxide, which aggravates the greenhouse effect. Green growing things normally offset global warming by sucking up carbon dioxide, but nitrogen on farm fields plus methane from decomposing vegetation make every farmed acre, like every acre of Los Angeles freeway, a net contributor to global warming. Fertilization is equally worrisome. Rainfall and irrigation water inevitably washes the nitrogen from fields to creeks and streams, which flows into rivers, which floods into the ocean. This explains why the Mississippi River, which drains the nation's Corn Belt, is an environmental catastrophe. The nitrogen fertilizes artificially large blooms of algae that in growing suck all the oxygen from the water, a condition biologists call anoxia, which means "oxygen-depleted." Here there's no need to calculate long-term effects, because life in such places has no long term: everything dies immediately. The Mississippi River's heavily fertilized effluvia has created a dead zone in the Gulf of Mexico the size of New Jersey.

America's biggest crop, grain corn, is completely unpalatable. It is raw material for an industry that manufactures food substitutes. Likewise, you can't eat unprocessed wheat. You certainly can't eat hay. You can eat unprocessed soybeans, but mostly we don't. These four crops cover 82 percent of American cropland. Agriculture in this country is not about food; it's about commodities that require the outlay of still more energy to *become* food. The grinding, milling, wetting, drying, and baking of a breakfast cereal requires about four calories of energy for every calorie of food energy it produces. A two-pound bag of breakfast cereal burns the energy of a half-gallon of gasoline in its making. All together the food-processing industry in the United States uses about ten calories of fossil-fuel energy for every calorie of food energy it produces.

That number does not include the fuel used in transporting the food from the factory to a store near you, or the fuel used by millions of people driving to thousands of super discount stores on the edge of town, where the land is cheap. It appears, however, that the corn cycle is about to come full circle. If a bipartisan coalition of farm-state lawmakers has their way—and it appears they will—we will soon buy gasoline containing twice as much fuel alcohol as it does now. Fuel alcohol already ranks second as a use for processed corn in the United States, just behind corn sweeteners. According to one set of calculations, we spend more calories of fossil-fuel energy making ethanol than we gain from it. The Department of Agriculture says the ratio is closer to a gallon and a quart of ethanol for every gallon of fossil fuel we invest. The USDA calls this a bargain, because gasohol is a "clean fuel." This claim to cleanness is in dispute at the tailpipe level, and it certainly ignores the dead zone in the Gulf of Mexico, pesticide pollution, and the haze of global gases gathering over every farm field. Nor does this claim cover clean conscience; some still might be unsettled knowing that our SUVs' demands for fuel compete with the poor's demand for grain.

Green eaters, especially vegetarians, advocate eating low on the food chain, a simple matter of energy flow. Eating a carrot gives the diner all that carrot's energy, but feeding carrots to a chicken, then eating the chicken, reduces the energy by a factor of ten. The chicken wastes some energy, stores some as feathers, bones, and other inedibles, and uses most of it just to live long enough to be eaten. As a rough rule of thumb, that factor of ten applies to each level up the food chain, which is why some fish, such as tuna, can be a horror in all of this. Tuna is a secondary predator, meaning it not only doesn't eat plants but eats other fish that themselves eat other fish, adding a zero to the multiplier each notch up, easily a hundred times, more like a thousand times less efficient than eating a plant.

This is fine as far as it goes, but the vegetarian's case can break down on some details. On the moral issues, vegetarians claim their habits are kinder to animals, though it is difficult to see how wiping out 99 percent of wildlife's habitat, as farming has done in Iowa, is a kindness. In rural Michigan, for example, the potato farmers have a peculiar tactic for dealing with the predations of whitetail deer. They gut-shoot them with small-bore rifles, in hopes the deer will limp off to the woods and die where they won't stink up the potato fields.

Animal rights aside, vegetarians can lose the edge in the energy argument by eating processed food, with its ten calories of fossil energy for every calorie of food energy produced. The question, then, is: Does eating processed food such as soy burger or soy milk cancel the energy benefits of vegetarianism, which is to say, can I eat my lamb chops in peace? Maybe. If I've done my due diligence, I will have found out that the particular lamb I am eating was both local and grass-fed, two factors that of course greatly reduce the embedded energy in a meal. I know of ranches here in Montana, for instance, where sheep eat native grass under closely controlled circumstances—no farming, no plows, no corn, no nitrogen. Assets have not been stripped. I can't eat the grass directly. This can go on. There are little niches like this in the system. Each person's individual charge is to find such niches.

Chances are, though, any meat eater will come out on the short end of this argument, especially in the United States. Take the case of beef. Cattle are grazers, so in theory could live like the grass-fed lamb. Some cattle cultures—those of South America and Mexico, for example—have perfected wonderful cuisines based on grass-fed beef. This is not our habit in the United States, and it is simply a matter of habit. Eighty percent of the grain the United States produces goes to livestock. Seventy-eight percent of all of our beef comes from feed lots, where the cattle eat grain, mostly corn and wheat. So do most of our hogs and chickens. The cattle spend their adult lives packed shoulder to shoulder in a space not much bigger than their bodies, up to their knees in shit, being stuffed with grain and a constant stream of antibiotics to prevent the disease this sort of confinement invariably engenders. The manure is rich in nitrogen and once provided a farm's fertilizer. The feedlots, however, are now far removed from farm fields, so it is simply not "efficient" to haul it to cornfields. It is waste. It exhales methane, a global-warming gas. It pollutes streams. It takes thirty-five calories of fossil fuel to make a calorie of beef this way; sixty-eight to make one calorie of pork.

Still, these livestock do something we can't. They convert grain's carbohydrates to high-quality protein. All well and good, except that per capita protein production in the United States is about double what an average adult needs per day. Excess cannot be stored as protein in the human body but is simply converted to fat. This is the end result of a factory-farm system that appears as a living, continental-scale monument to Rube Goldberg, a black-mass remake of the loaves-and-fishes miracle. Prairie's productivity is lost for grain, grain's productivity is lost in livestock, livestock's protein is lost to human fat—all federally subsidized for about $15 billion a year, two-thirds of which goes directly to only two crops, corn and wheat.

This explains why energy expert David Pimentel is so worried that the rest of the world will adopt America's methods. He should be, because the rest of the world is. Mexico now feeds 45 percent of its grain to livestock, up from 5 percent in 1960. Egypt went from 3 percent to 31 percent in the same period, and China, with a sixth of the world's population, has gone from 8 percent to 26 percent. All of these places have poor people who could use the grain, but they can't afford it.

Food is politics. That being the case, I voted twice in 2002. The day after Election Day, in a truly dismal mood, I climbed the mountain behind my house and found a small herd of elk grazing native grasses in the morning sunlight. My respect for these creatures over the years has become great enough that on that morning I did not hesitate but went straight to my job, which was to rack a shell and drop one cow elk, my household's annual protein supply. I voted with my weapon of choice—an act not all that uncommon in this world, largely, I think, as a result of the way we grow food. I can see why it is catching on. Such a vote has a certain satisfying heft and finality about it. My particular bit of violence, though, is more satisfying, I think, than the rest of the globe's ordinary political mayhem. I used a rifle to opt out of an insane system. I killed, but then so did you when you bought that package of burger, even when you bought that package of tofu burger. I killed, then the rest of those elk went on, as did the grasses, the birds, the trees, the coyotes, mountain lions, and bugs, the fundamental productivity of an intact natural system, all of it went on.

A FOREST'S LAST STAND

BARBARA KINGSOLVER

Xmul. X'pujil. Once you learn to pronounce the *X* as a "Shh...," the place-names of the Mayas sound like so many whispered secrets. So does the Mayan language that is still spoken, with quiet ubiquity, in the Yucatán. Along the rural roadsides, where fathers and sons walk in early light to the milpas, you can hear it. At the Merida market where women sit and lean their heads together behind stacks of tomatoes and *chaya* leaves, this language of secrets is passed along.

Leading south from the colonial city of Merida to the ruins of ancient Uxmal is an old road that rises into dry hills of farms and woodlands. This was the road we chose. As Steven drove, I navigated, using a map that showed a Mesoamerican culture's famous antiquities while somehow neglecting to mention that the culture itself was still completely alive. This was Mayan countryside. Nearly every little town had an *X* to its name, and every woman who walked along the roadside had on the Mayan dress, a lace-trimmed white cotton tunic brilliantly embroidered at the bodice and hem. All of the dresses were different, like eye-popping snowflakes, and they obviously weren't put on for tourists—we were by this time well out of tourist terrain. The women wore them when they did their marketing, laundering, and garden work, and even, as I saw once, when they fed the family hogs; miraculously, the dresses always seemed to remain dazzlingly white. To my eye, this was magical realism.

Our journey's end lay much farther to the south, in the humid forests that touch the Guatemalan border, but I tend to travel toward destinations the same way I look up words in the dictionary, getting sidetracked by every possible item of interest along the way. So we made a detour to inspect one of the notable antiquities on our map: Uxmal. Older by centuries than the aggressively heroic pyramids at Chichén Itzá, Uxmal's structures are just as tall but somehow less high-and-mighty. (They are also far less frequently visited by tourists, because they aren't as handily reached from Cancún.) The Pyramid of the Magician is round-shouldered and delicate—if the latter word can reasonably be used of a pyramid. Also the plaza at Uxmal is less grandly paved, more mossy underfoot than Chichén Itzá's. The soft ground swallowed the sounds of our steps as we walked through the enormous, silent city. Everywhere we looked, the facades were etched with turtles, monkeys, and jaguars, and the staircase guarded by feather-headed serpents. Living iguanas the size of small alligators perched on the cornerstones, glaring at us in a good imitation of the glowering stone heads above them. The limestone rain gods at Uxmal have looked down their huge, up-curled noses for nearly two thousand years, but the iguanas have less patience with the enterprise; they're inclined to roll off their posts and undulate across weathered steps. Winding paths lead out from the central plaza through the forest to other clearings and buildings: temples, ball courts, stories carved in stone. The fringe of surrounding jungle hides dozens more structures that have been left unrestored, consigned to crumble quietly under blankets of vine and strangler fig, keeping their secrets to themselves. As we walked the forest path, a light rain began to darken the gods' stone pates, imperceptibly dissolving their limestone, carrying off another small measure of history.

As we left the modern settlement that surrounds Uxmal, we recalled the counsel of friends in Merida not to head south into the sparsely populated state of Campeche without a full tank of gas. Adventurous but not foolish, we backtracked to the nearest PEMEX. Steven negotiated for *sin plomo* while I attacked the windshield, facing up to an omelette of Mexican insect life. I was mostly still lost in Uxmal's iguana dreams from the day before, but suddenly as I scraped at the windshield I found my attention snagged on a gigantic agrichemical-company ad painted on the building across the street. It showed a merry campesino dousing his corn with a backpack sprayer as huge green letters loomed in the sky above him: *Psst... Psst... There goes your security!*

Something must be getting lost here (or gained) in my translation, I thought; this was just too much truth in advertising. If Mexico is the NAFTA sister with the brightly embroidered dress and the hibiscus behind her ear, she is also the one whose reputation has been most tarnished by chemical dependency. The fields here are dumping grounds for DDT and virtually anything else ever deemed too toxic for U.S. consumers. The country's capital has our planet's most chronically poisoned air; the 1994 earthquake, in momentarily shaking traffic to a halt, afforded many of Mexico City's residents their first-ever sight of blue sky. The chemicals sign might just as well say *Psst... There goes your tourist dollar!* It's true that Mexico's siren song of beaches and margaritas calls out seductively to students on spring break, but North American travelers looking for nature unspoiled generally skip right over it to Costa Rica and points south.

In doing so, they fly directly over the place we were now setting out to look for on our full tank of gas, a putative gem of undefiled Mexico. Most people would be surprised to learn that the largest tropical forest on our continent stands on the southernmost reach of Yucatán and northern Guatemala.

By late afternoon we had reached its frontier, the Calakmul Biosphere Reserve. The tiny town of X'pujil ("Shh...pujil") guards a bend in the road and the ruins of Becán, a city even more ancient than Uxmal and even less visited by tourists. We stopped to walk Becán's tree-filled plazas and secret stone passageways in the company of no other humans at all, only birds. A turquoise-browed motmot peered down at us from a branch, its long tail clicking back and forth like the pendulum of a clock.

My map showed that the reserve pinched to a wasp-waist here at X'pujil. The wilderness broadens out as it stretches north to the Puuc Hills and far south into Belize and the Guatemalan highlands. In its southern reaches, this same forest shadows the ancient Mayan cities of Tikal and Uaxactun. The road we were on skirted the forest's edge. The sun had set, but the trees' upper branches were still lit like candles, aflame with birds. Keel-billed toucans hailed us from high overhead with huge beaks that looked freshly painted by an artist on a binge. In the treetops they threw back their heads and laughed their good-nights. An enormous lineated woodpecker's vermilion crest stood straight up as if frozen in fright. We rolled down our windows and breathed in rarefied steam. The boughs of a gumbo-limbo tree drooped low with roosting chachalacas, dark, chicken-size birds renowned for their remarkable singing style. But by now it was too late in the day for singing. Eyes in shining pairs blinked from the roadside: foxes, agoutis, maybe wild cats. The Calakmul Reserve is home to jaguarundis, ocelots, margays, pumas, and jaguars. It also conceals tapirs and opossums, turkeys colored like peacocks, orchids and bromeliads the size of turkeys, monkeys that hoot like owls, and owls with eyes in the back of their heads. Where the map shows a vast green emptiness, the land is alive.

That this wilderness still exists is something of an accident of geology. In this century the industries of deforestation moved south through Mexico like General Sherman, sweeping and burning it clear of its subtropical forests. The march came this far and then literally dried up. Even though the rains come heavy at times here, no streams at all cross the Yucatán's limestone surface; the same water scarcity that plagued the ancient Mayans has daunted modern ranchers in the region, preventing the successful development of this land for large-scale meat farming. So the Calakmul, for now, still belongs to jaguars and toucans.

Of course, birds and beasts alone have no power to save a Mexican forest. What the Mayans of old worshiped as gods, the modern Mayans tend to eat. Likewise the Chol, Tzeltal, and other groups fleeing Guatemalan repression and Mexican poverty. Some fifteen thousand refugees poured into this region in the mid-1990s. Their tradition of slash-and-burn farming demanded that they leave behind used-up cornfields every three years to clear new patches of forest. The Mexican government had designated the Calakmul forest a biosphere reserve in 1989, but signs posted to that effect tended to be read by the homeless as invitations to settle. "*Hooray*," refugees must have exclaimed at the sight of the forest preserve signs. "Nobody's living here who will give us trouble."

And so Mexico's last great forest, having held its own against timber magnates and hamburger franchises, seemed doomed to fall one branch at a time to cook fires and corn patches. But because of an extraordinary program launched in 1991, it still stands. In the villages surrounding the Calakmul another whisper was going around, maybe a feather of hope for the place. That was what we were looking for.

In the village of Nueva Vida, or "New Life," Carmen Salgado waved happily from her gate and invited us into her backyard garden. We told her we'd been sent from the *consejo*. Just north of the ruins at X'pujil we'd found the small concrete-block building that housed the farmers' co-op office, and the kind folks there had directed us to Nueva Vida. They'd promised that here and in the region's other small villages we might find an intriguing update on the civilizations we had been admiring in postmortem condition to the north, where the great pyramids poked out of the Yucatán forests. Here and now, in a cooperative of thirty-six families, papaya and lime trees shaded thatched houses elegantly constructed of smooth wooden poles. I kept studying them until the connection registered: These high-peaked roofs perfectly echoed the shape of the vaulted ceilings we'd seen inside every Mayan ruin we had visited. The architecture had preserved its central elements for thousands of years.

But here was another story—the village of New Life was looking very much alive. Outside Carmen's high-peaked thatched house, in her sunny garden, I stepped carefully to avoid solid plantings of cilantro, lettuce, and *chaya*—which she explained was a high-protein leaf crop that had been grown in the area since ancient times. A vine she called "nescafé" curled its tendrils around the wire fence that contained her compost pile; from its beans she made a coffee substitute and protein-enriched bread. We walked from her back gate down the gravel path through the village center, where a lush community citrus orchard offered oranges and grapefruits. A turkey paused to eye us, then continued stalking the ground under the citrus trees with a fierce forager's eye, taking seriously his job as the DDT of a new generation.

No snaking backpack sprayers will *psssst* in this Garden of Eden. Carmen informed us in no uncertain terms that chemical pesticides and fertilizers are beyond the means of the subsistence farmers here—and what's more, they are learning not to want them. Instead, they demoralize pests with a concoction of soap, onions, and garlic. Their reliance on organic methods of pest control and soil amendment allows these farmers self-sufficiency, while also ensuring that their notoriously poor tropical soil will improve with each crop, rather than deteriorate.

Carmen's broad, handsome face lit up as she explained these things. Although she has had almost no formal education, she is astute, articulate, and comfortable with visitors, a natural spokesperson for Nueva Vida and its new program. She grew up in one place and another in the poorest parts of rural Monterrey, without family, land, or much hope until she came here. She was lucky: She arrived just as a new environmental appreciation was dawning over the Calakmul forest, and with it a new approach to its conservation. Everything depends on these villages immediately surrounding the forest preserve. Nueva Vida is one of the seventy-two *ejidos,* or cooperative farms, that ring the Calakmul reserve in a protective belt, established by land grants assigned to groups of refugee families that otherwise, inevitably, would have consumed the forest from the inside out. The plan the reserve's managers came up with may seem contradictory to U.S. notions of wilderness preservation, but here in the land of the Maya it may just be the only right solution: Rather than fight a losing battle to keep people out, they would help them move *into* the forest. Recognizing that human habitation was an ancient and integral part of this ecosystem, the managers hoped that nature might best be preserved here by human residents who had a good enough reason to care for it. A boundary of settlements could buffer the forest against waves of outsiders moving farther in. The program's goal was to encourage these farmers to shift their long-standing war against trees into a peaceful coexistence.

But having a land grant means staying in one place and learning to call it home, no small departure for the refugee populations of Nueva Vida and the other *ejidos,* who previously spent their lives using up land and moving on. The concept of composting may seem obvious enough to the sedentary, but for those with no cultural memory of standing still for more than three years, seeing soil improve and fruit trees grow is a kind of miracle. It's almost impossible to explain what a huge leap of faith is involved here, even in a citrus orchard. I was at first amused and then, as I began to understand, profoundly impressed by the enthusiasm Carmen and her fellow *ejidarios* displayed for their orchards and gardens and even their simple, beautifully functional composting toilets.

Watching things grow, improving a piece of land—for historical refugee populations these are cultural accomplishments even more significant than learning to read and write or earning a degree. They embody a complete psychological transformation.

The transition has happened gradually, in daily lessons that come through patience and careful scrutiny. Don Domingo Hernández, an elder statesman in the neighboring collective of Valentín Gómez Faríaz, had a lot to tell us about that. He walked us out to his cornfield, where he was experimenting with soil-boosting cover crops, and gave us a lively lecture on the benefits of chemical-free agriculture: healthy soil microbes, nitrogen fixation, humus, conservation of moisture. Don Domingo tipped back his weathered cowboy hat, bent to scoop up a handful of black dirt, and held it out to me as reverently as any true believer might handle a relic of his faith. "Three years in this patch," he said, "and this is the best corn crop I've ever had. Next year will be even better."

The prime mover behind this change was not government charity but education provided by Pronatura, a Mexican conservation group, in concert with the U.S.-based Nature Conservancy and other private organizations. Working from a thatch-roofed office just north of X'pujil—which, like its demonstration garden, is open to visitors—a handful of agronomists and engineers offer ideas and technical advice; they are enormously respected by the region's farmers. Carmen was animated about this point. "Doña Norma came out from the *consejo* office and said to us, 'What are you wasting time for? Plant trees!' So we planted trees." This and dozens of other projects have given families like Carmen's a sense of belonging to their land, and a reason to stay. They have dug cisterns to catch rainwater from their roofs, following the instructions of a Pronatura engineer; they have also begun new beekeeping enterprises. A course in medicinal plants offered at the *consejo* office teaches women how to collect, process, and label all the remedies their families will need for infections and minor ailments. (The medicines are stored in plastic film canisters donated by conservation groups in the city.) After Hurricane Roxana devastated the southern Yucatán, when fevers and infections were scathing the peninsula, Carmen's collective had kilos of medicine on hand to donate to the relief effort.

"It's a much better life we have now," Carmen insists. "We were skeptical at first, and some still hold to the old ways. But really, the old way was that we ate rice and beans and drank coffee. The rice and coffee, we had to buy with cash. We have a healthier diet now with all the things we grow, and it's nicer, more interesting. Our kids like it better—that's how you know a change is going to stick." She gave my belly a glance, smiling at my incipient pregnancy. "For the kids, there is no going back; this is the life they will choose."

The stalwart tropical day had slipped away into evening as we were talking, and we now stepped outside Carmen's cool thatched house to watch the full moon rise. Orchids planted in tin cans bloomed pale and fragrant in the dusk. Normally they grow in the forest canopy, unseen by human eyes; Carmen called these her *huérfanas pobrecitas*, or "poor little orphan girls," because she'd salvaged them from trees that the men of a neighboring *ejido* had felled for lumber. Every collective includes arable fields and a parcel of forestland extending into the Calakmul reserve, to be used as the cooperative sees fit. Some are cutting their trees, sustainably, in a managed forestry program, but increasingly, others are not cutting at all. Carmen's group of women voted against clearing their twenty-five hectares for a cornfield, deciding that the parcel would be more valuable to them as it stood, since it provides flowers year-round for beekeeping as well as an inexhaustible apothecary. It's also a balm for the spirit. Carmen made us pause in our conversation to look at the moon, a perfect orange lantern cradled in the arms of a cecropia tree.

"Listen!" she commanded, her eyes bright. From the forest's edge a warm wind carried the scent of wild spices and the sweet call of a pygmy owl. Somewhere within the jungle nearby a blue-eyed jaguar crouched, searching the wind for signs of its age-old forest companion, the human animal.

Many kilometers from the bordering ring of villages, deep in the very heart of the reserve, the giant pyramids of the Calakmul ruins rest in permanent peace. At each turn of our journey we'd seen more remote, unvisited ruins, but this was literally the end of the road—the very edge of North America, beyond which no human residence or enterprise was to be found for a far cry. Our new friends from the *ejido* and *consejo* office had roused us in the early-morning darkness from the small thatched house on stilts where we'd spent the night, guiding us down the long, bumpy dirt road into the forest's heart with promises of the most dramatic sunrise of our lives. Now we groped our way by flashlight up deeply weathered steps to the top of the tallest pyramid. Mayan glyphs silently held their accounts beneath the industry of foraging ants. As the limestone softly crumbles, the forest retrieves it.

On a small platform atop a pyramid that was probably used for just such dramatic ceremonies in antiquity, our little group waited for the sunrise. I stood up, a little dizzy from the height—we were way above the treetops—and out of breath from the steep climb. I put my hand on my belly, where I carried the daughter whose name and gender I didn't yet know, and I whispered to my child internally: *Remember this with me. Once upon a time we were here, at the center of the world.* As far as I could see in every direction, a dark green sea of untouched forest rolled out to the whole encircling horizon. In a lifetime—mine, anyway—one is given this blessing only rarely: the chance to stand on high ground, turn in every direction, and see absolutely not one single sign of visible humanity. This is how the world once was, without our outsize dreams and dominion. Nothing surrounded us but the dark embrace of trees, except where the predawn light touched the eroded stone face of another pyramid rising above the canopy. Our friends pointed out a bump on the southern horizon that they said was the pyramid of Mirador, on the Guatemalan border. From here to there, when the sun was just right, a person could flash signals with a mirror; and from Mirador, someone else could signal farther south to Tikal, and so on, to the edge of the Mayan world. We stood at its very center. Then, between one held breath and the next, the sun appeared to us, scarlet and full-skirted on the horizon.

Very suddenly we found ourselves surrounded not by eerie silence but by a wilderness of wake-up calls. A troop of howler monkeys began to stir in the treetops just below us, letting loose a loud, primordial bellow. Emerald battalions of parrots darted past in formation, flashing in the whitewashed light.

Then came the chachalacas, the chickenlike birds we'd seen the previous day, whose call, I had been promised, I would never forget. "*Sh...!*" our friends said, "*Escuche,*" and we listened, but I didn't hear it at all.

And then I did: a barely audible chorus in the far distance, Cha-*chalac*? Quietly, distantly, their neighbors answered back, *Cha*-chalac! The more I listened, the more plainly I could hear how they followed the call-and-response rhythm of a gospel choir: Cha-*chalac*? *Cha*-chalac! They stirred one another to voice in increasing numbers to announce their revelation. This forest, I began to understand with a chill, was entirely filled with chachalacas. The birds themselves don't move, but their song does as they awaken one another each morning, their dawn chorale moving through the whole jungle in a vast oratory wave. The rising tide of their gospel song raced toward us, growing louder, louder and faster: Cha-*Chalac*? *CHA*-CHALAC! CHA-CHALAC! *Glory hallelujah*! The song came from everywhere at once, a musical roar like water, and then like water it divided, passing around us as a rush of singing, and then it receded and fell away—*Cha-chalac!*—toward the southern horizon. Finally it faded out of earshot.

None of us spoke. I imagined this wave of hallelujah traveling all the way to Guatemala and beyond, on down to the southern edge of the jungle, where the trees once again gave way to roads and cornfields, billboards and gas stations. But we were still deep inside a green, crowded world where parrots and monkeys were not isolated survivors but citizens of a population. It was a city of animals here, as surely as each mute temple stood for a city of people who had once carved their reverence for animals in stone and climbed up to greet the dawn.

Of course, they aren't gone, the Mayans. On carved slabs of stone they left us clear pictures of their world, with man and beast facing off nose to nose in a thousand configurations: warrior and monkey; jaguar and emperor. The Mayans' ways and reverences have endured like stone, altered through seasons of sun and rain. Now in our latter days, iguanas scowl at tourists, and farmers may raise up great clouds of death on the vermin, but sometimes another story can root itself and take hold. In some quarters, farmers named Carmen and Don Domingo rule, in a reign that allows no poison and holds its breath for the moon and smiles at the sweet nightsong of an owl. Human and beast together may persist in this place, as they have always done, since the days when God was a feather-headed serpent.

THE FARMER AS CONSERVATIONIST?

LAURA L. JACKSON

Edward Abbey was a writer of sublime wilderness essays, a spiritual leader of the conservation biology movement who did much to protect the American West. Abbey was lyrical, righteously angry, hard-drinking, and hilariously irreverent. He also seemed not to know or care where his beer came from. Like Abbey, early conservation biologists focused on wilderness, while taking for granted the ecological sacrifice zones where their food (and beer) were produced. Fortunately, this oversight is being corrected, and now many scholarly journals and conference presentations do address agricultural landscapes: how can agricultural landscapes be designed to provide habitat for native pollinators and birds? what modifications in the livestock industry would allow it to coexist with healthy populations of large carnivores?

I live and study in the great American Corn Belt, the vast ecological wasteland of northern Iowa, so this shift in attitude is important to me. Just a tenth of a percent of our original tallgrass prairies remain, mostly in remnants smaller than my city block. I can drive to either border of the state and see little but crops during the whole trip, the monotony broken only by thin strips of streamside woodlands. The livestock are invisible, known by their housing—rows of large metal buildings—and their smell. Because over half of the grains are fed to livestock, either here or in other states or abroad, one might call this the "protein production landscape." But it is also the high-fructose corn syrup landscape, the ethanol and biodiesel landscape, and the food additive landscape. Unfortunately, it is a model of agriculture emulated around the world.

The soils and climate are too perfect and the land prices too high to restore significant landscapes to tallgrass prairie. So for the last dozen years I have been starting many of my lectures with the question, "If conservation biologists could have anything they wanted, what would agriculture look like?" I and many others have conjured ideal designs for agricultural systems, prescribing conditions and practices that would maximize sustainability and minimize off-site environmental impacts. Organic farming, wild corridors, buffer strips along streams, crop diversification, extended crop rotations, rotational grazing, agroforestry, perennial grain crops—there are many good ideas for what farming in any given place "ought" to look like. The response to this advice has been an earsplitting silence. Conservation biologists and other advocates of biological diversity, as it turns out, have no say whatsoever in the design of the agricultural system.

Who does make the important decisions in the agricultural landscape, and how are they constrained? To whom should our advice be directed? This is similar to the dilemma faced by an unfortunate restaurant-goer who, after waiting hungrily for forty-five minutes, receives a burned and incorrect order from a clearly exhausted server. Yell at the waiter or waitress, or ask to see the manager?

Is the farmer—a land owner, embedded in cultural tradition and skills, armed with scientific research, served by technological advancements, and guided by his or her personal ethic—the designer of this landscape? Or is the landscape designed by something larger, by what might be called the industrial food system? That is, a small group of agricultural input suppliers, grain handlers, and livestock processors, global corporations that shape U.S. farm policy, public university research, and Americans' eating habits?

The sustainable agriculture movement has had a longer history than conservation biology in trying to improve agriculture, and thus we might learn something from it. I have had the opportunity to observe the sustainable agriculture movement since the mid-1970s when my parents founded The Land Institute in Kansas. Since moving to Iowa in 1993 I have become a member (nonfarming) of both The Land Stewardship Project (MN) and the Practical Farmers of Iowa and, since 2003, have served on the advisory board of The Leopold Center for Sustainable Agriculture. The views and any errors expressed here are strictly my own, and do not represent those of the Leopold Center or my own institution, the University of Northern Iowa.

The Leopold Center in particular occupies a precarious perch, housed in the epicenter of industrial agriculture, Iowa State University (ISU). Literally dependent on Iowa farmers who pay fertilizer and pesticide taxes, this institution's dilemmas are instructive and serve to reveal broader truths. Recently, the Center and its home institution have arrived at an impasse over its emphasis and direction. In fall 2005, the Dean of the College of Agriculture removed the director of the Leopold Center, citing widespread and longstanding criticism among its important constituents. At the heart of the impasse was, I believe, the question of whether substantial environmental and social problems in the Corn Belt could be solved by simple technical improvements in farming, or a radical design overhaul. In other words, should we speak to the waitress or the manager?

HISTORICAL ROOTS: TARGETING THE FARMER

The Leopold Center for Sustainable Agriculture was created as part of the 1987 Iowa Groundwater Protection Act, landmark legislation that taxed fertilizer and nitrogen sales to fund public education and sustainable agriculture research. The current budget of the Center is about $1.5 million, with $1.1 million coming from the state and the rest from grants. Thus the Center is dependent on the good will of the State Legislature and on the farmers who buy inputs for corn and soybeans.

The language of the legislation indicates how controversial it was. Sustainable agriculture was defined as "the appropriate use of crop and livestock systems and agricultural inputs supporting those activities which maintain economic and social viability while preserving the high productivity and quality of Iowa's land." The role of the Center was to

(1) Conduct and sponsor research to identify and reduce the negative environmental and socioeconomic impacts of agricultural practices; (2) Research and assist in developing emerging alternative practices that are consistent with a sustainable agriculture; and (3) Develop in association with the Iowa cooperative extension service an educational framework to inform the agricultural community and the general public of its findings.

The emphasis on economic viability and high productivity was a conscious effort to distance this project from the urban environmentalists, back-to-the-land purists, nostalgia-mongers, and Greenpeace type activists who had been criticizing mainstream agriculture but had no idea where their food came from. Note also that this mandate and charge was *not* going to regulate farming practices. Change would come about by good science and farmer education. Landowner ethics would be shaped by new knowledge of the negative impacts of certain practices. The legislation, strongly influenced by Aldo Leopold's 1949 essay "The Land Ethic," resonated with many Iowans' sense of personal responsibility.

THE FIRST 10 YEARS: SUCCESS AND CREDIBILITY

The Leopold Center became a catalyst for on-farm research in association with members of a new organization called the Practical Farmers of Iowa (PFI). Despite early struggles to establish credibility, the Leopold Center was successful in forming productive collaborations between farmers with ideas for cutting input costs and ISU researchers. For agricultural scientists, research on real farms, within the whole farming system instead of isolated on experiment stations, became respectable for the first time.

A principal focus was on reducing the input costs of farming. Studies found ways to reduce tillage, reduce pesticide use, improve weed control, protect soil quality, and improve the productivity and profit margin of pastures through new rotational grazing techniques.

Related to reducing input costs was limiting the loss of nutrients from fields. Water quality in Iowa is impaired by both surface runoff of sediments and phosphorus, and underground leaching of nitrogen. A successful initiative established wide buffers of perennial vegetation along streams. The buffers harbored wildlife and stopped sediments and phosphorus from getting into the waterways. To address nitrogen leaching (a leading cause of shallow groundwater pollution and a growing Dead Zone in the Gulf of Mexico), several studies looked at how to apply nitrogen fertilizer more efficiently. Abundant nitrogen is required for top corn yields, but it has a tendency to leach from the soil before the crop gets a chance to use it. Researchers found that they could increase efficiency by testing for the amount of nitrogen in the soil before applying it, and by applying it in two smaller doses in the spring instead of a single dose in the fall. Statisticians even detected a significant down-tick in per-acre nitrogen use, as the research information got out to farmers.

A third major area of research activity at the Leopold Center involved investigations into alternative buildings for hog production. Year in and year out hogs were the farm's most profitable "mortgage burner." In recent years production had shifted from many, medium-sized operations with outdoor facilities, to fewer, larger operations with indoor "confinement" systems. Hog confinements allowed for faster weight gains and leaner pork, but were criticized for their odor, spill-prone liquid waste handling systems, and inhumane conditions. The "hoop building," a Swedish-designed alternative to the metal and concrete confinement buildings had most of the advantages of conventional confinements, but cost much less to build. Open on both ends, they provided daylight and natural ventilation. Animals stood on straw bedding over a soft manure pack instead of concrete or metal slat floors over liquid manure pits. Animal waste was handled as a composted solid, cutting odor and eliminating the risk of a manure spill. The pigs had room to move around, play, build nests, and socialize according to their natural instincts, reducing the incidence of stress-related disease and cannibalism.

These project areas received widespread support from mainstream agriculture. They were producing results that the average cash grain farmer or hog producer could conceivably apply. Taxpayers were getting their money's worth, and the legitimacy of the Leopold Center was high.

INTRACTABLE PROBLEMS : KNOWING WHEN TO SAY "ENOUGH"

Somewhere in the mid-1990s, however, a number of problems began piling up at the doorstep of sustainable agriculture research—problems that did not submit to farm-oriented solutions. First and foremost, farmers kept going out of business. Despite their best efforts, innovation, increased efficiency, and university-led research, farmers found that it was simply not profitable to cut input costs and raise more environmentally sound crops and livestock. Their products may have been more sustainably produced, but they were still often sold at below the cost of production, dumped into the same bin as everyone else's grain, and slaughtered alongside everyone else's livestock. The price was identical, whether they took care of their land or not. (In the case of hogs, mid-sized farmers were actually paid less for the same quality animals because they could not promise delivery of at least one semi-trailer load per week.) No matter how clever farmers had become in cutting down on fertilizer and pesticide use, the conventional market still did not pay them for their enhanced land ethic.

Excess nitrogen was a second intractable problem. It was rising steadily in surface waters and shallow wells every decade, and the Dead Zone in the Gulf of Mexico continued to expand. Leopold Center-sponsored research had found ways to reduce losses of nitrogen from fields by a few percentage points, but for many cash grain farmers with a thousand acres to plant in the spring, it was not worth the extra risk and management costs. The late spring soil testing had to be done when the farmer was busiest and weather most unpredictable. The window of time for planting and spraying was too narrow to fool around with split nitrogen applications, especially when the product was cheap. Nitrogen fertilizer was treated as an inexpensive form of crop insurance, so farmers made sure they had "plenty." In a good year, corn supplied with plenty of N will make top yields. If the weather or weed control were less than ideal, leading to lower-than-expected yields, the extra nitrogen didn't cost much. In addition, fewer farmers had access to alternative sources of fertility (manure, forage legumes) because fewer could raise cattle profitably. Across the state, nitrogen application rates started to rise again.

Soil scientists and agronomists across the Midwest turned to other methods of reducing nitrogen losses—minimum tillage systems, buffer strips, different kinds of drainage, chemicals to stop the soil microbes from converting ammonia to nitrate, scrubber wetlands, tractors guided by satellites, even trenches filled with sawdust. These techniques either had very minor and variable impacts on nitrogen loss, or were too expensive and labor intensive to implement. Researchers found that corn-bean cropping systems on tile-drained, prairie soils could leak as much as 60 percent of the applied nitrogen.

Numerous studies reiterated that reducing the acreage of row crops by rotating them with alfalfa, oats, and pasture grasses could reduce N losses by 97 percent or more. For significant change to happen—not just for the best farmers on flat land, but for the average farmers on sloping land—we would need for soil to be covered with vegetation and filled with live plant roots for six to nine months of the year instead of three. Perennial plants tie up slippery nitrogen into plant organic matter and feed the soil microflora, building soil health and protecting water quality. In short, we needed to perennialize the Midwestern landscape with cover crops, hay, and pasture. That meant bringing the livestock back from the distant feedlots and out of their confinement buildings, out onto pastures, back to the farm.

The collapse of an Iowa tradition, the independent family hog farm, was another clear signal that adopting low-input practices on the farm was not enough. The breaking point came in the winter of 1998. Large, confinement feeding operations had long-term contracts with meat packers, so they were somewhat insulated from swings in market prices. In February 1998 hog prices on the shrinking open market dropped to $6 per hundred pounds ($35 was considered break-even). By the end of that winter, only two groups of pork producers were left standing: large conventional confinement operations holding contracts with companies like Heartland Pork and Swine Graphics, and the "hoop group," marketing humanely raised, antibiotic-free hogs to gourmet restaurants in San Francisco.

There was a growing realization in the sustainable agriculture movement that federal farm policy, particularly commodity support payments, largely determined farmers' choices. And farm policy was formed with heavy participation from agribusiness and the food industry. A series of studies by Mary Hendrickson and William Heffernan revealed an increasing dominance of just a few corporations in control of all sectors of the food system: crop and livestock genetics, grain milling and processing, livestock feeding and slaughter, secondary processing, and even retail. The top four firms in each sector controlled from 40 percent to as much as 80 percent of the total market, seriously reducing the competitiveness of those markets.

Ultimately, many scientists across the country, members of the Leopold Center staff, some Leopold board members, and many others nationwide began to recognize a kind of defeat: the corn-bean-concentrated livestock system was artificially supported by government payments, fundamentally flawed ecologically, and ultimately, beyond meaningful help. We had run out of technical fixes for two crops that had miraculous yields, but serious ecological limitations. The more radical elements within the sustainable agriculture movement insisted that conventional conservation efforts such as no-till farming and buffer strips, while laudable, just merely enabled a bad system to persist a little bit longer. Conservation programs to help farmers build terraces and grass waterways, protect highly redouble land (CRP) and wetlands (WRP), while effective, were beginning to look like distractions and decoys.

It was time to stop asking farmers to do more. Even the most virtuous, conscientious farmers could not fix this. Like the servers in a poorly managed restaurant, they could not be held entirely accountable for systemic failures.

THE TURNING POINT: LANDSCAPE DESIGN OVERHAUL

In response to these breakdowns, the Leopold Center established three focal "Initiatives" in Policy, Marketing and Food Systems, and Ecology in 2000. Requests for proposals in the Ecology Initiative sought projects that would capture the power of "free ecosystem services" such as natural pest control from diverse cropping systems. Research and demonstration projects favored perennial grass, such as rotationally grazed dairy and grass-fed beef. Grant proposals that focused on improving ("tweaking") corn, soybean, or livestock production in conventional systems were rated less favorably than those that involved greater plant cover, more grass, or a combination of the two.

The Marketing and Food Systems Initiative helped farmers coordinate with processors, distributors, and retailers. In collaboration with the Practical Farmers of Iowa and other foundations (notably the W. K. Kellogg Foundation) they turned attention to promoting and developing markets for food with a regional identity and food produced in sustainable ways. They began collaborating with schools of business and big food distribution companies like Sysco, and working on "equitable value chains" so that farmers would no longer be price-takers but equal members of a value chain in which each segment received a fair share of the profits.

The Leopold Center was not alone in this shift; around the Midwest, sustainable agriculture organizations such as Practical Farmers of Iowa, the Land Stewardship Project, and the Institute for Agriculture and Trade Policy increasingly moved in the direction of food systems. It was a new world: organic markets were growing at close to 20 percent per year and eco-labels, farmers' markets, and community-supported agriculture (CSA) enterprises were taking off across the country.

Meanwhile, intensive row crop farming and confined livestock kicked into an even higher gear. The 1996 Farm Bill as it was eventually implemented separated government payments from farmers' compliance with conservation plans or production limits. Production of corn and soybeans expanded into the former wheat belt (Kansas and North Dakota). Market prices hit new lows, spurring more hog and chicken confinements, a bonanza of new corn ethanol plants (also due to more federal and state subsidies), increasing land and rent prices, and a commodity farm program that costs U.S. taxpayers about $20 billion per year. Another cause of intensification was biotechnology. Round-up Ready soybean genetics made weed control easier, and soybean acreage rapidly increased by 1.3 million acres (15 percent), making Iowa's landscape even more homogenous.

So as industrial agriculture kept intensifying, the Leopold Center turned to providing for its own separate markets. Rather than move down a common road and argue which direction it should go, there were now two separate roads, running in nearly opposite directions.

THE REACTION AND AN ANALYSIS

The upshot of these actions by the Leopold Center was a growing tide of resentment by individual farmers and most certainly by commodity groups (*e.g.*, American Corn Growers Association, National Pork Producers Council), the American Farm Bureau Federation, and powerful agribusiness organizations. The dean of the College of Agriculture at Iowa State reported receiving many communications from sources considered to be important constituents of ISU, all deeply concerned over the direction of the Center. While bestowing praise on the Marketing and Food Systems program, which has enjoyed great success, the dean and others felt that it was still "niche" marketing, irrelevant to the average farmer.

That stung, but it was true. Local food programs do bring economic opportunities to farmers. But according to calculations by Mimi Wagner at Iowa State University, the state's population could be fed on roughly 880,000 acres, or 3.5 percent of the land in farms in Iowa. As the dean asked pointedly, "What are we going to do with the other 24 million acres?" (There are roughly 25 million acres in row crops in Iowa.) Organic, local, and specialty markets for sustainably produced products, though growing fast, are still too small to occupy the vast expanses of the North American prairies, at least under the current cheap energy regime. There is hope that highly differentiated food products—memorable, high quality, sustainably produced foods with a story behind their production that is in synch with consumer values—could go mainstream and eventually command enough acres to make a difference at the ecosystem level. The Leopold Center has taken the lead by organizing an "Agriculture of the Middle" task force to help organize regional networks of mid-size farmers. It may yet be successful, especially if American consumers continue to put their food money where their values lie. Just how much land could be affected by this trend has not yet been determined.

The dream I once found eminently reasonable—to return to a modern version of the pre-1950 landscape of extended crop rotations, rotational grazing, extensive hay and pasture, and reintegration of crops and livestock—is likewise considered unreasonable even by those sympathetic with the basic idea. The entire infrastructure is gone: the fences, sale barns, packing plants, the know-how, the young people willing to get up every morning and do chores (the average Iowa farmer is 62 years old), to stay at home all winter, not to mention the markets. Maybe a few operations here and there can manage to do this. But again, "What are we going to do with the other 24 million acres?"

Eventually, the Leopold Center director was removed and retitled a Distinguished Fellow. A recent external review recommends "blurring the line between mainstream and sustainable" as well as a rapprochement with conventional stakeholders and strategic partnerships with commodity groups (corn, soy, hogs). I believe that ultimately the Center's course correction was rejected because it violated a foundational myth of the mainstream agricultural community, the myth that the farmer is in charge, responsible, and can overcome everything, even markets controlled by just four firms; even the inherent leakiness of the corn/bean cropping system; even the conflict between the *nature* of domestic animals and their role as protein machines in industrial-scale production. Quite a nostalgic view, considering the source.

The industrial food system has every reason to perpetuate the nostalgic myth of farmer heroism and responsibility for the land (for a good example, see BestFoodNation.org). By doing so, it can distance itself from responsibility for environmental degradation caused by agriculture and avoid more expensive regulations or, far worse, a much-needed overhaul of the basic structure of industrial agriculture.

A MODEST PROPOSAL: IT IS TIME TO SPEAK TO THE MANAGER

So far, we have avoided the unpleasantness of this option by urging a boycott of the restaurant. Certainly, the movement to "vote with your food dollars," has its merits and has certainly gotten the attention of mainstream food corporations, who are hurriedly buying up smaller organic food companies. But no consumers' movement has succeeded with a boycott alone. Buying local, organic, grass-fed, predator-friendly, cage-free, shade-grown, dolphin and salmon-safe foods is part of the solution, but it is difficult to imagine it will make much of a dent in the cheap, fast, and convenient food system that sculpts the character of the Midwest. We need a parallel strategy.

Let me suggest that we work to convince the food and agriculture industry to embrace goals for rural economic viability and ecosystem health as well as food production. (1) At least 80 percent of farms should make a profit 8 out of every 10 years without government commodity payments. (2) Reduce farm and food sector consumption of fossil fuels by 50 percent. (3) Reduce the nitrates and sediment at the mouth of the Mississippi River by 10 percent every decade.

(4) Leave land in grassland or forest sufficient to support viable populations of a significant carnivore such as the bobcat. The means to achieving these goals would be left to the giant agribusinesses who control nearly every aspect of agricultural production. The welfare of farmers must be protected by insisting that environmental quality and biodiversity targets be coupled with specific goals for improved socioeconomic health in rural areas.

I do not pretend to know how those goals would be met. I do not know whether putting political pressure on the food industry would lead to greater justice for the farmer. But I do know this is a discussion we need to have—a discussion with the management and the owners for a change, not just the waitresses and the waiters. The most powerful members of the value chain, those closest to the farmer (inputs, grain milling, meat packing, and so on) should be asked to figure out ways to achieve these societal performance goals. They, in turn, will be forced to create more meaningful partnerships with farmers and rural communities, because they will not be able to meet those goals otherwise. In this way, farmers and rural communities—and the land community to which they belong—can gain power within the value chain that starts with soil and ends with the evening meal.

In the future, agribusinesses will need to take responsibility for more than just a "safe, inexpensive, and abundant food supply." This sounds idealistic, but consider recent decisions within the apparel industry to assume greater responsibility for working conditions in the small factories across the developing world. Brands like Nike and Gap have had to respond to stakeholder concerns, developing protocols and monitoring to improve the human rights in the independently owned shops that supply their products. The alternative to ethical responsibility on the part of any entity in society is regulation. But if it comes, that regulation should be distributed in relation to the power of the various partners in the food value chain, not to the farmer alone. Conservation biologists, by asking to speak to the manager and owner of our industrial food system instead of offending the wait staff, will have taken a critical step in improving the health and biological diversity of the agricultural landscape.

Section II

CORE ISSUES

There is no such thing—so the saying goes—as a "free lunch." Previous assumptions that the natural world could withstand a limitless onslaught of industrial activities are now being cast in a harsh light. Certainly crops need irrigation to thrive; but when aquifers are depleted, groundwater poisoned, and entire river systems dewatered, somewhere along the line there will be hell to pay. Fish need healthy creeks, streams, rivers, and lakes, as do nearly all species at some point in their life cycles. Monocultures, seemingly invincible in their sheer production power, have proven imminently vulnerable. That fragility is becoming blatantly apparent even in the synapses of the farm system, such as with pollinators that shuttle genetic material from flower to flower and make much of agriculture possible in the first place.

Farming and ranching methods that support the ongoing survival of wild nature are taking on many forms and evolving as agriculturalists assess and apply conservation knowledge. Nearly a century ago, the organic movement, one of the primary catalysts for this necessary marriage of conservation and agriculture, began challenging the application of industrial logic to farming and livestock husbandry. Our present and future solutions may indeed demand inspiration and an evolution of wisdom from the past. Diverse cropping systems provide natural resilience to pest and disease pressures, in contrast to entire valleys and plains planted with a single crop type and plant variety. With an attitude toward coexistence rather than eradication, even watershed-altering species such as beavers may become accepted and perhaps well appreciated. The same can be said for predators. For centuries, Europeans have managed livestock amid greater wolf populations than many North American ranchers experience. If one single shift carries the potential for change at a continental scale, it is the restoration of grasslands, and with them, a turn toward perennial-based agriculture.

SALMON STAKES

TED WILLIAMS

Last fall the Bush administration proved what everyone else already knew: Fish need water.

In an effort to appease irate irrigators, the U.S. Bureau of Reclamation (BuRec) had dewatered the Klamath River, which drains a 9,691-square-mile watershed of high desert, woods, and wetlands in southern Oregon and northern California. By July the agency had cut the flow from its Iron Gate Dam from 1,000 cubic feet per second—previously deemed by the administration as the bare minimum necessary to prevent extinction of the system's coho salmon—to about 650 cfs. From July 12 to August 31, 2002, more water went down the main diversion canal to irrigators than down the river to salmon.

Meanwhile, farmers were getting—and wasting—so much water that they were flooding highways and disrupting traffic.

State fisheries biologists, commercial fishermen, sport fishermen, Klamath Basin Indian tribes, and environmental groups had repeatedly warned the Bush administration that such dewatering would devastate chinook salmon and steelhead trout populations and perhaps usher cohos into oblivion. After the National Marine Fisheries Service (NMFS) determined that BuRec's plan would indeed jeopardize the existence of coho salmon, the leader of the NMFS team writing the biological opinion (required by the Endangered Species Act when a federal action might affect a listed species) says he was ordered to change his finding and that, when he refused, his superiors made the changes themselves.

In mid-September, four months into BuRec's new 10-year water-distribution plan, chinooks, cohos, and steelheads from the icy Pacific hit the low, warm, deoxygenated river and turned belly up. The mortality estimate was 33,000 fish, mostly chinooks. From all reports, it was the largest die-off of adult salmon ever. Bright, robust fish, many over 30 pounds, covered gravel bars, blocking foot traffic, fouling the water, filling the air with a stench you could taste.

The Klamath system was once the nation's third biggest producer of Pacific salmon. All five species flourished there, as did steelhead, green sturgeon, and two species of native mullet known locally as the Lost River sucker and the shortnose sucker. Now chinooks and steelheads are down from their presettlement abundance by something like 90 percent. Sockeye, pink, and chum salmon are extinct in the basin. Cohos are listed as threatened. Until the 1970s the Klamath tribe caught thousands of pounds of mullet; now it takes one fish a year for ceremonial purposes. Both mullet species are endangered, and the green sturgeon is being considered for listing. These fish have been flickering out because BuRec's 95-year-old Klamath Project has replumbed the Klamath system with a network of 6 dams, 185 miles of canals, 516 miles of lateral ditches, and 45 pumping stations. Now water flows everywhere it never belonged.

About 280,000 of the basin's original 350,000 acres of wetlands and shallow lakes have been drained or filled. Still, the Klamath Basin—a.k.a. "Everglades West"—provides refuge for 80 percent of all waterfowl that negotiate the Pacific Flyway. In winter these birds help sustain the largest population of bald eagles in the contiguous states. The U.S. Fish and Wildlife Service operates six national wildlife refuges in the basin. But, like BuRec, the service is part of the Department of the Interior; and, under the Bush administration, it has become part of the problem.

Klamath Basin farms get about 12 inches of rain and 100 growing days a year. Before there were crop surpluses, water shortages, and endangered species, it seemed a dandy idea to make this high desert bloom. These days it's an insane waste of money and resources—like transporting iron ore by air. Until October 31, 2002, when *The Wall Street Journal* ferreted it out, the Bush administration had been suppressing a peer-reviewed U.S. Geological Survey (USGS) study that found that agriculture in the Klamath Basin generates $100 million a year compared with the $800 million generated by recreation, such as camping, boating, rafting, swimming, and fishing, and that restoring water to the river would boost this last figure to $3 billion. The study also determined that buying out the farms and protecting the land would create $36 billion in benefits at a cost of $5 billion. In an internal USGS memo obtained by the *Journal,* an agency scientist revealed that the regional director "wants to slow [release of the study] down because of high sensitivity in the Dept. right now resulting from the recent fish kill in the Klamath. Suffice it to say that this is not a good time to be handing out this document."

In the Klamath Basin the government gets farmed a lot more than the land. There is scant demand for most of the crops grown; sometimes they're even plowed back into the ground. Originally it cost farmers nothing to get a permanent irrigation hookup to BuRec's public-financed Klamath Project. Now, on top of this, they get electricity to operate irrigation pumps at one-sixteenth of fair market value, a lower rate than their ancestors paid in 1917. During the dry summer and fall of 2001, basin farmers—some irrigating normally with emergency wells drilled at public expense—harvested $48.6 million in state and federal relief. Many reported their most profitable year ever.

Those who did best didn't own land; they leased it from the Fish and Wildlife Service at $1 per acre while reaping a minimum of $129 per acre in farm subsidies. They farmed the Lower Klamath and Tule Lake refuges, supposedly devoted to waterfowl and (in the case of the latter) bald eagles, which depend on waterfowl. These two refuges, once the flagships of the refuge system, are now national embarrassments. Of America's 540 national wildlife refuges, they are the only two that permit commercial agriculture. The farming program, administered on 25,600 acres, requires about 60,000 acre-feet of Klamath River water per year; pollutes river and wetlands with phosphates and nitrates; and loads land and water with pesticides, including two neurotoxins, 14 endocrine disrupters, and 11 carcinogens. When water is scarce, as it usually is in the basin, marshes go dry so farmers can get water. Waterfowl have plummeted from 6 million or 7 million in the 1960s to about 1 million today.

On both refuges the Fish and Wildlife Service is in gross violation of the National Wildlife Refuge Improvement Act of 1997, which stipulates that permitted activities be "compatible with the major purposes for which such areas were established." The service attempts to justify its farming-first policy with the Kuchel Act of 1964, which permits agriculture in national wildlife refuges. But the statute requires that such agriculture be consistent with fish and wildlife management. After dewatering, polluting, and poisoning marshes and river, farmers produce potatoes and onions and grain—far less nutritious to waterfowl than wetland plants.

For anyone still in doubt, the summer of 2001 proved that there isn't enough water in the Klamath River for fish, waterfowl, and agriculture. Something had to give; that was agriculture and refuges. The river's endangered mullet and threatened coho salmon had first dibs on water. Then came tribal-trust resources—mainly chinook salmon. If there was any water to spare, it could go to agriculture and refuges. That's what state water law and the Endangered Species Act said.

But politics said otherwise. Incited and assisted by property-rights groups, irrigators organized a "bucket brigade." (This was modeled after the Jarbidge Shovel Brigade of Elko County, Nevada, which fantasized that it had "sovereignty" over federal lands and, on July 4, 2000, hacked an illegal road through the Humboldt-Toiyabe National Forest and habitat of the threatened bull trout.) On May 7, 2001, some 15,000 farmers, politicians, and property-rights activists (many bused in by the Farm Bureau) scooped buckets of water from a lake that feeds the Klamath River and passed them hand to hand through downtown Klamath Falls and into an irrigation ditch. The media circus attracted politicians, who puffed and blew about the evils of the Endangered Species Act. Senator Gordon Smith (R-OR) vowed to introduce a bill that would "reform" the act. "We must never feel it's okay to say that sucker fish are more valuable than the farm family," he proclaimed. Representative Wally Herger (R-CA) called the situation a "poster child" for Endangered Species Act reform. And Representative

Greg Walden (R-OR) lamented that surgery on the act had to start with another "dust bowl." Later, an organizer of the Jarbidge Shovel Brigade arrived with a 10-foot-tall bucket. *The Pioneer Press,* a local weekly, started a "virtual bucket brigade" by e-mail, in which 70,000 people expressed support for the irrigators.

On June 29 an irrigation-canal headgate was illegally opened and water released from Upper Klamath Lake. BuRec shut it. Twice more an angry mob, now encamped, opened the gate, and twice more BuRec shut it. On July 4 about 150 demonstrators formed a human chain, shielding vandals who cut off the headgate's new lock with a diamond-bladed chainsaw and a cutting torch. The sheriff announced that he wouldn't bust anyone, because they were only "trying to save their lives." A deputy drove up in his cruiser, lights flashing, removed his hat, and replaced it with a farmer's. With that, he opined that Oregon environmentalists were likely to elicit such violence as "homicides." He even suggested two potential victims: Andy Kerr and Wendell Wood of the Oregon Natural Resources Council. Finally, BuRec called in U.S. marshals.

Two weeks later Senator Smith's amendment to the Interior appropriations bill—it would have required federal agencies operating in the Klamath Basin to ignore the Endangered Species Act, legal obligations to Indian tribes, and the Clean Water Act—failed by a vote of 52 to 48.

On July 24 Interior Secretary Gale Norton divined that there was water to spare in Upper Klamath Lake and ordered 75,000 acre-feet released to farmers. "Unfortunately," she declared, "none of this water will reach the national wildlife refuges because there simply is not enough water to do more than provide a little relief to some desperate farm families during the remainder of this season." She went on to suggest that wintering bald eagles could be artificially fed.

The drought was much less severe in 2002. Still, with great fanfare, the Bush administration cut off minimum flows to fish, tribes, and refuges in order to provide irrigators with full deliveries. On March 29 the headgate at Klamath Falls was again opened—this time by Agriculture Secretary Ann Veneman and Interior Secretary Gale Norton, who were on hand to emcee the ceremony. Veneman spoke of the administration's "commitment to help farmers and ranchers recover from losses suffered last year." Norton gushed about how nice it was to be "providing water to farmers."

Diverting so much of the Klamath for irrigation required brand-new science. Under the mandate of the Endangered Species Act (ESA), the NMFS had issued a biological opinion that such dewatering would jeopardize coho salmon, and the U.S. Fish and Wildlife Service had issued a biological opinion that it would jeopardize the mullet. So Norton asked the National Research Council (an offshoot of the National Academy of Sciences) to review the two documents. In February 2002, after just three months, the NRC panel hatched an interim draft report, alleging that the biological opinions weren't supported by enough science. Armed with this opinion, the president's Klamath advisory team (consisting of the secretaries of Interior, Commerce, and Agriculture, and the chairman of the President's Council on Environmental Quality) ordered new findings from the NMFS and the Fish and Wildlife Service.

On October 28, 2002, Michael Kelly, the NMFS biologist assigned to write the biological opinion, filed a federal whistle-blower disclosure with the U.S. Office of Special Counsel, charging that the team's recommendations for minimum flows were twice rejected under "political pressure." His main complaint was that the required analysis for the Reasonable and Prudent Alternative—the part of a biological opinion that tells an agency (in this case, BuRec) what it should do to avoid jeopardizing a listed species and which, in this case, had been suggested by BuRec—was intentionally not carried out, and that a specific risk to coho salmon that he and his colleagues had identified had been intentionally ignored.

A month after his disclosure, in his first interview with the media, Kelly told me this: "We were ordered to interpret the NRC report as recommending that the Bureau of Reclamation could avoid jeopardy by operating as it had for the previous 10 years. But simple logic and a basic understanding of the Endangered Species Act regulations can demonstrate that any 'recommendation' in the NRC report does not make sense in an ESA context. One of the problems we have with the NRC report is that the panel never defined what kind of confidence they wanted. We biologists felt like they were a bunch of PhD's accustomed to reviewing peer-reviewed scientific-journal articles that require a very high level of confidence. A biological opinion is not that kind of document. The regulations say you use the best information available. You have to make a conclusion. And when you're unsure, you give the benefit of the doubt to the species."

Most whistle-blowers put up with lawbreaking until late in their careers, when they haven't got much to lose. But Kelly is only 37 and has a young family to support. When the Public Employees for Environmental Responsibility came to his defense, it warned him that if he blew the whistle, he might get lucky and hang on to his job, but that he should pretty much expect to lose it. I asked Kelly if blowing the whistle had been worth the risk. "They [his team's superiors] did a masterful job of forcing us to the point where I just couldn't participate any longer," he said. "The only way for me to continue would have been to violate the Endangered Species Act. I just couldn't do that. I wouldn't want to be continually participating in such egregious rule breaking and mismanagement of resources. In the past there was always subtle political pressure. I'd hear a supervisor say, 'Well, we can't recommend that under this administration.' It was de facto pressure. But this was finally something that was so blatant I had to say something."

Since Kelly's disclosure, two Oregon State University researchers who had been investigating the NRC document—fisheries professor Douglas Markle and graduate student Michael Cooperman—have reported that it is riddled with errors, such as incorrect water-quality data, faulty fish-population models, selective use of data to support "a conclusion they had already reached," and even reference to nonexistent species.

Enemies of the ESA and the press framed the controversy as a choice between fish and farmers. "The Bush administration knew exactly what side they wanted to be on—the side of the farmers," says Mike Daulton, Audubon's assistant director of government relations. "So, dismissing the opinion of the NMFS and others and disregarding the downstream tribal fishermen, they decided to put on paper this 10-year plan to basically guarantee flows to irrigators."

There is only one solution to the Klamath water crisis: End lease farming on the refuges and buy farms and water rights from willing sellers. Before the summer of 2002 the federal government was committed to just this. But it gave up when it ran into fierce resistance from business interests that profit from farming, such as pesticide and fertilizer distributors, and from farmers who lease land cheaply on the refuges and therefore profit from subsidies. In an October 23, 2001, letter to Representative Wally Herger, the Tulelake Growers Association tried to get Phil Norton, who was then manager of the Klamath refuges, disciplined for alleged violations of the Hatch Act, which proscribes lobbying by federal employees. As evidence the association cited comments attributed by the media to Phil Norton, such as: "We are trying to fix the system so that it works again, but there's a lot of land that, frankly, never should have been put into agriculture production."

Last June the Fish and Wildlife Service, which is controlled by the Secretary of the Interior, completely reversed itself, issuing a Finding of No Significant Impact from farming in the Tule Lake National Wildlife Refuge. And despite a 94 percent favorable response in the public-comment period, the service rescinded its 1999 ruling that irrigation on the refuges would be permitted only in years when there was enough water to sustain wetlands. It abandoned its buyout effort. No longer did refuge spokespeople say that lease farming on the refuges "had to go." Instead, they proclaimed that farming was "compatible" with their mission.

The Klamath Water Users Association prevailed on Representative Greg Walden to kill an amendment to the 2002 Farm Bill that would have provided $175 million to buy farmland from willing sellers in the Klamath Basin. This so infuriated farmers who own land and have long favored a buyout that 50 of them wrote the association as follows: "To prevent this unfortunate situation from reoccurring and to prevent any future legal action, we request that all future association activities purporting to represent Klamath Basin Water Users on any major issues, such as retirement of land, be submitted to a vote of the landowners prior to any public announcement or official position statement."

Among the signers was John Anderson, 50, who runs beef cattle and grows a few crops on 3,500 acres in Tulelake, California. The drought of 2001 hurt him badly, wiping out 100 of 150 acres of peppermint and making him even more determined to get into a business more practical and profitable than trying to make the desert bloom.

"The buyout has become an emotional issue that has built on itself," he says. "Logic has been lost. People go around saying, 'By God, we're not going to let the government take it,' and 'These environmentalists are full of bull.' I'd say more than 50 percent of the farmland is available for federal buyout right now." A lot of landowners aren't talking because they've been intimidated by property-rights barkers. Anderson, who is not among them, says he has received death threats by phone in the middle of the night.

On October 2, 2002, after salmon had been dying in the lower river for two weeks, the Pacific Coast Federation of Fishermen's Associations, the Earthjustice Legal Defense Fund, the Wilderness Society, Trout Unlimited, the Yurok tribe, and Representative Mike Thompson (D-CA) held a press conference outside the Interior Department building in Washington, DC, to announce their lawsuit against BuRec and the NMFS for violating the Endangered Species Act. Thompson had the Yuroks ship out 500 pounds of dead salmon with which he and his fellow plaintiffs festooned the park across from the Interior building. So rancid was the shipment that Federal Express at first refused to deliver it. "It was amazing how quickly the flies found those fish," recalls the Wilderness Society's Pete Rafle. "I now understand why the theory of spontaneous generation held sway. I've got a pair of shoes that I'm going to have to resole or burn. I wasn't expecting puddles."

"I think there's been a real lack of understanding that the salmon are connected with the farming practices," Representative Thompson told me. "Unless you know the area, you don't necessarily know that the two are connected, and that's been a big problem. So it has come down to God-fearing farmers versus hippies and fish. That's not what it's about at all. It's about livelihoods in the lower basin."

The only people to express surprise at the fish kill worked for the Interior Department. Steve Williams, director of the Fish and Wildlife Service, showed up at the press conference to lament the plaintiffs' "premature rush to judgment" and proclaim that it was "too soon to draw conclusions" about what might have killed the salmon—roughly the equivalent of a parachute manufacturer suggesting that sky divers scraped from asphalt might have died on the way down from food poisoning.

Sue Ellen Wooldridge, Gale Norton's deputy chief of staff, asserted that the government can't release much water from Upper Klamath Lake because of the endangered mullet, failing to mention that if it hadn't diverted the river for full deliveries to irrigators in violation of the Endangered Species Act, there would have been more than enough water for mullet, salmon, and refuges.

Finally, James Connaughton, chairman of the President's Council on Environmental Quality, offered this explanation: "There will always be setbacks because we don't have an ultimate authority on how natural systems work. The trick is to manage risk in a way that minimizes and localizes and creates limited opportunities of time for those setbacks to occur."

In other words, the president's top environmental adviser expects the public to dismiss what's apparently the biggest salmon kill in history as just another bum hand in a game of five-card draw, played with the public's fish and wildlife as the ante. The Klamath tragedy isn't an isolated event. On September 30, when the salmon die-off was at its peak, the administration was giving away federal water a thousand miles east, on the Gunnison River in Colorado, thereby desiccating the Black Canyon of the Gunnison National Park and jeopardizing four endangered fish and a world-famous trout fishery. Earlier in the month Interior declined to appeal a bizarre court ruling that canceled the water right of Deer Flat National Wildlife Refuge in Idaho, a refuge dedicated to *waterfowl.* Since 1973, when the Endangered Species Act outlawed these kinds of risks, no other administration has been willing to take them. Now they're a habit with the Bush team, and it isn't winning any pots.

WHEN THE EPIDEMIC HIT THE KING OF CLONES

GARY PAUL NABHAN

& ANNA GUADALUPE VALENZUELA-ZAPATA

It is an old story remembered in the Zapata family, known by all of Ana's kin. When her older brother came home drunk, Lupe Zapata ran to hide amidst the corrals behind their house. She didn't hide for fear as much as for the pain it gave her to see her brother in such a drunken condition. She then swore to God: "Lord, bring some plague to dry up all the agaves, to rid us of all this [drunkenness]!" Sixty years after she cast her spell, that plague has come, reshaping the trajectory of the tequila industry as dramatically as any factor has within our lifetimes.

In 1988 we began to hear rumors from agave producers, rumors about a plague—based on several crop pathogens—more aggressive than anything they had previously witnessed. From that time on, we began to observe a blight that made one agave leaf after another turn yellow, encircled with rust-toned bands the *mescaleros* called *el anillo rojo.*

In some plants, the leaves would lose both their sky-blue intensity and their rigidity; they would roll up, and split open and ooze below the terminal spine, probably as the result of another pathogen. The putrefying remains of the dying plants were not unlike those suffering from a known fungal infection that occasionally affected stressed agaves, but the damage did not seem to be from a single cause.

When the plague came to the attention of our old friend Guadalupe Fonseca, who directs all fieldwork for one of the major agribusinesses in the region, he commented that it seemed as though plagues always arrive in the best times, when weather and business are favorable. We had all heard of the "gangrene" that agaves suffered during the first tequila boom a century before. And our colleagues in Oaxaca, Claudia López and Felipe Cruz, had been shown other analogous diseases there by local *magueyeros* just four years earlier. From that time on, we saw such infestations advance from one field to the next until it was evident in one fifth of all agaves planted in Jalisco, damaging nearly forty million plants in 1998 alone. From Oaxaca to Nayarit, there was a plague upon the land.

BECOMING BOTANICAL DETECTIVES

What we saw was exactly what other researchers had begun to document: a rotting of agave flesh caused perhaps by fungus infesting the stalks, caudex, and base of the plant. These fungal pathogens ruthlessly attacked the *cogollo,* or terminal growth bud. The fungus behaved as though it were a strain of *Fusarium,* a cause of wilt in many crops. But we also recognized that there was another element to the putrefaction, not unlike the symptoms seen during the last major agave disease outbreak. That epidemic had been caused by a bacterial stem rot known as *secazón* (the big withering). That round of damage was blamed on the bad guys known as *Erwinia,* a number of strains of bacterial pathogens found to wreak havoc on both wild and cultivated plant tissues. *Erwinia* bacteria often infect plants that are already stressed by freezes, drought, winds, or excessive heat, but they also do time after the plants have died, speeding up tissue decay as part of the recycling crew.

Regardless of how many pathogens were involved, there was something about this epidemic that was vaguely reminiscent of the plague that devastated agave crops during the first tequila boom of the nineteenth century. Curiously, in both cases, the epidemics occurred not long after a large quantity of vegetative offshoots were planted in monocultural stands on extensive acreages, in response to meteoric rises in tequila demand. We were of course aware of the theoretical risks associated with any monoculture, but we quickly realized that the even age of the transplanted clonal offshoots was also an aggravating factor. If the vast majority of agaves out in the fields were of just one age class, then their enemies—as well as their natural allies, perhaps—would have a field day. The problem was one of demographic vulnerability—two-thirds of the two hundred million agaves growing in Jalisco were planted within just a couple years of one another, all derived from the same clone, blue agave.

That bit of news meant that all the eggs had indeed been put in one basket. Under such circumstances, how could there be any capacity at all to be buffered genetically from an epidemic? This genetic vulnerability, which we first predicted several years before tragedy struck, has recently been confirmed by scientist Benjamín Rodríguez Garay, who explained it to science reporter Laura Romero: "The tequila agave casts its blue color on sixty thousand hectares of hills and valleys where not a single other kind of agave is allowed a foot-hold, and its means of propagation by cloning in such large quantities has caused a grave problem for the crop—that of genetic uniformity."

Although some cryptic somaclonal variants with disease resistance might occur within the two hundred million offshoots set out in the countryside, this kind of genetic variation has hardly been studied and would have been hard to mobilize in time to fend off the epidemic. Worse yet, the semi-domesticated progenitors and relatives of the blue agave clone have been virtually outlawed by the industry, and could hardly be located for resistance screening. The blue landscape that has been packed so full of agaves is now withering, threatened by its very success.

How can further losses be averted? How might health return to the land of tequila? How can we respond to the "economic disaster" that has already affected the thirty-five thousand families that rely on tequila production for their primary means of livelihood?

RETURN TO INTERCROPPING

Regardless of the level of genetic uniformity among the agaves themselves, there is less disease in virtually all tequila plantations that intersperse rows of peanuts or green beans between rows of agaves. These ground-covering, nitrogen-fixing legumes suppress the weeds that serve as vectors for diseases, and improve the soil. Ironically, the industry has discouraged intercropping over the last thirty years, considering it less efficient than agave monoculture as a use of arable lands suited to tequila cultivation. With the agave-culture tradition, intercropping was the rule, not the exception. Several of our own field studies have demonstrated the advantages, as well as the limitations, of intercropping agaves or other monocots with legumes.

RETURNING TO SPATIAL MIXTURES OF DIFFERENT AGAVE VARIETIES

It has been convincingly demonstrated, theoretically as well as in the field, that mixing resistant and susceptible varieties of a monocot crop in the same field, or in adjacent fields, slows down the spread and virulence of diseases and pests. In the case of the tequila-growing area of Jalisco, there was no doubt a mix of fiber and beverage agaves grown in the same area up until the time of the first boom, when blue agave clones were selected for their short maturation cycle. Since that time, several of the cultivated varieties with very distinct fiber characteristics and rosette architectures have been lost from the landscape as plantings of significant scale. However, several of these nearly extinct land races or heirloom varieties, including *zopilote* and *pata de mula*, have been recently rediscovered and given taxonomic rank.

These materials should be among the highest priorities for conservation *in situ* and *ex situ*, as well as for inclusion in agroecological experiments undertaken to solve current problems. Studies of their leaf morphology, fiber anatomy, pollen, and stomata are now underway at the Universidad de Guadalajara. At the Jardin Botanico de la UNAM and the Centro de Investigación y Asistencia en Tecnología y Diseño del Estado de Jalisco (CIATEJ), micropropagules of these cultivated varieties, as well as other agave species, are being screened for disease resistance to the bacterial and fungal pathogens implicated in the red ring epidemic. Those that have resistance can be regenerated through somatic embryogenesis in less than six months, and then rapidly multiplied by other tissue culture techniques. However, rather than planting a single resistant selection on large acreages, it would still be more prudent to plant intermixtures of various agave varieties to slow down the evolution of virulence in the fungal and bacterial pathogens. To achieve such a mosaic of plantings of various agave varieties, the industry would need to reconsider its exclusive use of the *azul* cultivar, and allow other varieties to be used in tequila production, just as it did a few decades ago.

RETURNING TO CROSS-POLLINATED AGAVE CROPS

Although many agaves are capable of vegetative propagation as well as sexual reproduction, few blue agave plants are ever allowed to flower, thus nectar-feeding bats and other floral visitors are unable to cross-pollinate them with pollen from nearby wild or cultivated plants. Bats such as the *murcielago magueyero (Leptonycteris curasoae)* have historically been responsible for the diversification of certain agave lineages, including that of *Agave angustifolia.* Cross-pollination often fosters genetic recombination in agaves, fueling the process of reticulate evolution that has allowed agaves to adapt to new abiotic stresses and to biotic challenges from pests and pathogens.

Over the long run, cross-pollination by bats or by human plant breeders will probably be necessary to keep the domesticated species of agaves evolving in response to new anthropogenic environments and stresses. However, it is doubtful whether allowing some tequila agaves to flower will produce the needed results in time to slow down the current epidemic. It will, however, recruit and sustain populations of now-declining bats that are responsible for pollinating a number of agave species along a fifteen-hundred-kilometer nectar corridor between Jalisco and the U.S.-Mexico borderlands. Restoring the relationship between bats and paniculate agaves makes good sense for agave germ plasm conservation in general, even if it does not immediately aid tequila growers.

EFFECTS OF THE ECONOMIC CYCLE ON AGAVES

The very fact that agaves have a long life cycle affects disease control strategies in many ways. Just as other crops are influenced by commodity price fluctuations, so are tequila crops, but in peculiar ways. Prices may fluctuate several times over the life span of a multi-annual crop like that of blue agave, which may take as much as twelve years to complete its entire life cycle. A good price at the time of harvest stimulates the expansion of plantings and the demand for farm workers. But if prices plummet due to declines in demand or to over-production, investments in the management and care of immature crops are reduced. Plantation managers pay laborers less to keep the agaves free of weeds, so pests and diseases are more likely to spread. When the *sobreoferta* or "excess supply" peaked between 1993 and 1998, some plantations were abandoned, left to die or to be overrun by weeds.

There are many adverse effects of these fluctuations on the regional economy. When prices are high, land owners wish to pack as many agaves onto their fields as possible, so that monoculture is favored over intercropping. As prices decline, the industry lacks the extra resources to establish intercrops on available parcels. There are similar effects on the quality and diversity of vegetative offshoots used for plantings. When prices are high, the demand for vegetatively propagated plantlets is also high, so that the clones that are most prolific are used to establish more plantations, regardless of their susceptibility to stresses. Poor quality offshoots get planted simply because demand creates a relative scarcity of carefully selected propagules. When prices drop, there is hardly any pressure to select only the highest quality propagules, because hardly anyone is planting at that time anyway.

In this manner, the economic cycle actually heightens the region's vulnerability to infestations by fungal and bacterial pathogens. In the early 1990s, prices for blue agave began to decline, and so the manpower dedicated to weeding and removing sick agaves from fields was diminished. Many fields were abandoned altogether, leaving a young cohort of plants on the land without plans for harvesting them. There were few incentives for investing in intercropping or seasonal grazing between rows, because the densest monocultures remained the most cost-efficient fields for harvesting. But the fields then became infested with pathogens that swept through, moving from plant to plant with utter ease. Meanwhile, the clone, in all its splendor of genetic uniformity, lacks the capacity to coevolve with the pathogens in a way that allows resistant clonal variants to persist.

As the 1990s came and went, the period of *sobreoferta* was followed by a shortfall of agaves relative to the global demand. Wildcat harvesters formed clandestine crews that entered abandoned fields at night, removing any agaves that were reaching maturity. Prices for *materia prima* tripled, since diseases had decimated even the most well-kempt plantations. For the first time in thirty years, some tequila firms' annual budgets were running in the red; their production costs had suddenly gone through the roof. There were twice as many distilleries as five years earlier, so that profits from the ever-rising demand for tequila were divided among many niche markets. To no one's surprise, investments in advertising and marketing reached an all-time high, as many of the older firms had to peddle harder simply to stay in place.

It is hoped that the Consejo Regulador del Tequila can help reduce the viciousness of such cycles, smoothing out the ups and downs of supply and demand. The trouble that remains is one of education: few economic botanists ever come to understand the economic cycles of agro-industries, and few agricultural economists ever come to understand how the life cycles of plants affect the dynamics that they study. Although the life span of an agave is long relative to the amount of time it takes for agave aficionados to obtain their undergraduate and graduate degrees, this dilemma is unlikely to be resolved within just one generation of professional development. The agaves are there, in the fields, waiting for us to get wise enough to help them out of this mess.

WILD WORK CREW

SCOTT McMILLION

Lynn Burton has a pretty good crew of farm hands. They labor all night without complaint. They ease his work-load and they put cash in his pocket. All they want in return is an occasional armload of aspen branches and some peace and quiet.

Long considered the bane of rural and suburban residents, beavers get more persecution than respect. They dam up culverts, flood fields, and topple beautiful trees, just because they just can't help doing the jobs nature has hard-wired them to perform: chewing wood and building dams. Burton, a biologist and range specialist for the Gallatin National Forest, says beavers get a bum rap. "They're the least understood, most undervalued animal in the world," Burton said. "They're the cheapest labor you'll ever find. I've never met a lazy beaver and they're trainable."

Training might be too strong a word. But Burton has shown he knows how to give the buck-toothed rodents some guidance and direction. Do it properly, he said, and they'll do a lot for you. A tour of his 18-acre property on the north bank of Rocky Creek east of Bozeman provided some impressive results. Since he bought the place two years ago, beavers have built six new dams. That created ponds, which hold

> There are plenty of tools available to help people live with beavers and to help reap their benefits while minimizing or avoiding the damage they can cause. "They're very useful and can be very destructive at the same time," said Mike Ross, a wildlife technician at the Montana Department of Fish, Wildlife and Parks who has worked extensively with beavers and their human neighbors, mostly to help people get rid of them. To keep beavers out of culverts, you can build a "beaver deceiver," a funny-looking fence that runs at an odd angle to the mouth of the culvert. Since beavers like to build dams at right angles to stream flow, a well-built fence will keep them away. While it might cost $400 to build, you'll pay that much each time a man with a backhoe shows up to unclog the culvert. Also, be careful how you install culverts. The sound of trickling water stimulates beavers to start building, so make sure the water moving through a culvert doesn't have much "drop" and can move silently.

water and dampen soils for a long stretch, the same way a sponge will become saturated even if only a corner of it is touching water. The high water table means Burton no longer has to irrigate his hay meadow, which saves on power bills and labor, and he still puts up enough hay to get his horses through the winter.

Plus, the beavers are improving habitat for fish, birds, and other wildlife. "If I was ever to sell this place, they're going to fall in love with this," he said, pushing through the thick creekside brush, a healthy trout stream at his side and birdsong filling the air. "They're not going to fall in love with my hayfield." Living with beavers isn't always easy and it takes some effort. Burton knows his property intimately, knows exactly what he wants the beavers to accomplish and where. If he wants them to dam a certain spot—say to slow the creek's flow and protect a sluffing bank—he'll put a few rocks in the stream to form a bed for a new beaver dam. The rodents do the rest. He decides which trees he wants to keep and which he wants the beavers to take. He wraps the keepers in chicken wire to deter the busy chewers. The rest of the time, he hopes the beavers fell the trees in the proper direction. "Sometimes they aren't very bright," he said of his field crew. So if it

You can protect valuable trees by wrapping their bases in chicken wire. Also, having a dog around the property can keep beavers out of your yard. If water is backing up too far behind a beaver dam, perforated pipe punched through the dam can allow some flow, and headgates can even be installed. Allowing beaver to live and do their work provides several benefits. The ponds and dams offer fish and wildlife habitat. They raise the water table, which reduces the need for irrigation, and they extend stream flows later in dry summers. They filter sediments and toxics from runoff. They hold back floodwaters, which cost the nation $4.2 billion annually. Each acre of wetland acts as a "safety valve" for 69 acres of uplands, biologist Lynn Burton has calculated. Burton's concept of trying to work with and manage beavers instead of trying to eradicate them is a good idea, Ross said, but an uncommon approach. "I'm impressed," he said of Burton's ideas. "The guy's smart."

looks like a tree is leaning the wrong way, Burton's not afraid to help things along with a chainsaw. And if the beavers get too busy; if, for example, they start flooding his pasture, he has nothing against dispatching a few of them with bullet or trap. He has no illusions about creating a wilderness. He's working on a managed landscape, but one that works with beavers instead of against them.

"It's the first I've heard of anybody doing anything like that," said Mike Ross, a wildlife technician at the Montana Department of Fish, Wildlife and Parks in Bozeman. "The only thing I've dealt with is the opposite." People who live along streams tend to get upset when a prized cottonwood or aspen suddenly becomes cordwood, and they want the sawyer removed. Since there are few places where beavers are welcomed after a trap and transfer, that usually means a death sentence. But killing beavers is a short-term solution. In a year or two, others will move in if the habitat is good and your problems might get even bigger. Most of the time, you're better off sticking with the beavers who know your property, Burton says. They're less likely than new arrivals to cause mischief because, like humans, when beavers find a new home, they quickly start changing things to their own liking.

Burton is already working with the Gallatin Wildlife Association, trying to spread the word about beavers: how to live with them and how to make them work for you. He hopes this fall to begin offering a series of classes and seminars on the topic. One of his neighbors, Pete Fay of Rocky Mountain Farm, is already a believer. "I've always really liked the beavers," Fay said. "I've had neighbors that hated them." When he first moved to the property in 1981, he said he was upset when beavers took out a big cottonwood tree. "My first reaction was, 'that stupid rodent killed that beautiful tree,'" he recalled. But then at least 100 new sprouts shot up to take the dead tree's place, stabilizing the bank and perpetuating the riparian forest. Fay said he realizes his productive vegetable and tree farm lies on ancient beaver ponds that silted in over the centuries. "I owe my five to seven feet of black soil to the beavers," he said.

Much of America stands on old beaver ponds. Before trapping began in the 16th century, there were up to 400 million beavers in North America. Biologists calculate that means there was a beaver pond every 500 to 2000 feet on almost every stream on the continent. Nearly all the beaver were wiped out by fur trappers by 1850, and homesteaders quickly claimed the fertile meadows that had been beaver ponds. They built roads and homes, installed bridges and culverts. But when beavers started coming back, the landowners began to feel invaded. Burton, for one, is putting out the welcome mat. He even hauls aspen branches—a beaver delicacy—to the creekbank. And he's not done yet. He figures he needs a couple more beaver dams on his stretch of creek. And he's hoping that will attract even more wildlife. "Wood ducks," he said. "I'd really like to have some wood ducks."

A GRASSLAND MANIFESTO

BECKY WEED

Small lessons around the homeplace are one way to wander into wild farming. But if one wonders about the fate of the homeplace, the journey can't end there. Either way, close or far, it's all about grass.

On a small scale, sheep ranching at home in Montana, we've watched coyotes watch us as we both mark territory. Mostly unwatched, mountain lions have harvested whitetail deer in the back fields and occasionally lambs in drought years. In the last few years, a herd of elk have made a regular circuit of pastures and brush in our neighborhood. In a tiny patch of woods on the neighbor's place, a bow-hunter watched a black bear wander right under her tree stand while she waited for those elk to pass through. All of this in spite of encroaching suburbia. None of these residents have put us out of the sheep business. They've tested and taxed us, but they've livened the journey for us and for our customers.

But our place is small, and the view necessarily expands, at first to the scale of a mountain range. The Bridger Mountains right behind us, it turns out, are one of the critical, vulnerable linkages in habitat for wolverine and other threatened species between the Rocky Mountains to the south, and the Belt Mountains and Canadian Rockies to the north. That realization led to queries outside of our farm. Indeed, we've learned of quiet families in the Rocky Mountain Front north of us managing ranches with grizzly bears; about ranchers in Alberta coexisting with wolves for decades; and mega-conservation buyers in Idaho revising grazing management in high desert and forest allotments to work toward ecological restoration goals on the scale of hundreds of thousands of acres.

There are even promoters of a bison ecology. Some are capitalist ranchers who love the merger of good business, beautiful prairie, and big steak. Others belong to interdisciplinary coalitions that envision a "wildlife economy" across millions of acres in the Northern Great Plains in the United States and Canada. Current residents, according to some scenarios, would be able to remain on their land in such an economy. Grassland restoration, with its full suite of participating residents, from bison to burrowing owls, would be the dominant regime.

All of these examples are of course anomalies in American agriculture and direct contradictions to "modern" industrialized agriculture. The notion of a restored bison ecology throughout vast portions of the Great Plains seems especially far-fetched, like something an enviro-on-hallucinogens or an eastern academic would dream about. (Indeed, they have.) Even if by some miracle wild farm ventures were to suddenly become the successful wave of the future, and all the major public and private western range operators were to suddenly withdraw from critical habitat and/or espouse wildlife-friendly practices, we still couldn't say that "Farming with the Wild" had "arrived" just because some sectors had arrived at wild farming. To understand why, a rancher must reluctantly explore brutal paths beyond the homeplace.

Within North America, western ranchers produce only a small fraction of the U.S. meat supply. Cowboys both real and imagined have comforted themselves with the western illusion of disproportionate political power, while their economic autonomy and clout have withered. Meanwhile, a dense cobweb straps them, along with their grain-farming neighbors and the abundant ranchers and farmers of Canada and the American Southeast, to the "other" landscape of North America—the Midwest. All are beholden, to varying degrees, to one dominant force emanating from the Midwest: artificially low feed-grain prices. Every strand of the cobweb (feeder steers trailering to auction and feedlot; feedlot effluent migrating to groundwater; ranch kids graduating to land grant universities or cities; boxed beef coalescing in fast food grinders) has been smothering the art of grassfarming, and obscuring the wisdom of wild grasslands for roughly half a century.

The meat *production* infrastructure engendered by such subsidized grains, that is, the corn-soybean-feedlot machine, has fostered a perverse price discovery process in which incentives for healthy meat are turned on their head, yielding antibiotic-laden, *e. coli*–contaminated, and hormone-infested beef. Perhaps most importantly, the "cheap" corn diet has fundamentally altered the fatty acid profile of America's dominant meat supply, degrading the health not only of domestic livestock, but also of the people who eat it.

The centralized meat *processing* infrastructure manufactured by the corn-soybean-feedlot machine has in turn disabled direct consumer ties to grass and growers on a massive scale. So now, even as the penetrating consequences of our national grain-fed dependence have begun to dawn on many growers and consumers, we wonder, is it possible to reverse the 20th century grass-fed to grain-fed transformation?

Compelling storytellers are building upon meticulous research outside the industrialized agricultural mainstream to publicize the interdisciplinary consequences of the feedlot machine: the effect of confined animal feeding on livestock and human health (Jo Robinson); the corn connection to obesity, diabetes, and other public heath quagmires (Michael Pollan); the sociological outcomes cultivated by centralized meat-packing and the fast-food industry, that is, occupational hazards, rural crime, low wage levels that perpetuate the demand for "cheap" food (Eric Schlosser); continental-scale ecological consequences of monocropping and excess nitrogen fertilizer applications (Richard Manning); and so on. These and other writers are inspiring consumer activism built on the power of the food dollar. Their call for conversion to grass-fed is exciting, empowering, and essential, but is it enough?

As a sheep rancher buried deep in the details of direct marketing, and sticky at the margins of the commodity cobweb, where can I turn to assuage my doubts?

- *Academia?* University ag research these days is largely beholden to corporate-driven, high-input husbandry/factory-suited genetics that are not geared to a grass-fed economy. Exceptions to this rule are not sufficiently powerful to induce industry-wide change on their own. But they could be important nonetheless.

- *Food Industries?* Our nation's cheap food dogma engenders cheap labor practices and narrow definitions of "efficiency," which in turn require cheap food policies. To oversimplify: if you live on box store wages, you probably have little alternative but to eat fast food. That vicious circle ensures that industry will not instigate reform of our current industrialized model without external pressure to do so. Mainstream livestock producers typically consider themselves to be part of the industrialized food system, so they're caught in the same vicious circle; they're not leading fundamental change.
- *Innovative Livestock Producers?* The most successful producers and direct marketers of grass-fed meat owe much of their commercial success to the essence of niche markets—scarcity and special stories. The incentives for preserving or at least passively accepting that scarcity act as powerful disincentives for food system transformation (or ecosystem-scale environmental protection) by even the progressive industry faithful.
- *Environmentalists?* The conservation community is intermittently effective at articulating the livestock "problem" but cannot be expected to, on its own, take on the bloody, profit-needy, inglorious task of working with domesticated animals and their handlers to achieve an alternative system.

No simple saviors here. The skeptic can find sobering evidence that, so far, an understanding of grass-fed meat production and its repercussions remains marginal in four key places: American medicine, land grant universities, USDA policies, and the bulk of Midwestern farmland. While a wellness newsletter from Berkeley may tout the benefits of grass-fed meat, the vast horsepower of pharmaceutical companies is churning on the income of cholesterol-lowering drugs, a manufactured antidote to our diet of *grain*-fed meat. A small handful of researchers explore the benefits of re-perennializing the upper Mississippi drainage, but the lion's share of agricultural resources is perpetuating sod-busting crop systems. We pour carbon sequestration research dollars into tweaking tillage practices in grain fields, instead of restoring grass. We even export our perverse dependence on corn and soybeans by quashing more diversified agriculture overseas, indirectly through domestic commodity subsidies, and directly, by forcing markets for genetically engineered seed. Consumer activism is mighty, but can buying habits change all this? No one can do it alone. And no alliance can do it without a vision for agriculture.

Even at the heart of progressive agriculture, within the booming organic movement, there is evidence of a faltering, or at least incompletely articulated, vision. The current angst and divisiveness between the small-scale/local/organic vs. industrialized/globalized organic sector says less about conflicts in peoples' motives than it does about uncertainty in their vision. The concerns are diverse—pesticides in produce, hormones in groundwater, the ethical welfare of chickens, the threats to open space, the genetic integrity of native plants, the livelihoods of indigenous peoples across the globe...what a mouthful. Such concerns are driving wonderfully positive, but *ad hoc*, changes in agriculture. When does the availability of marvelous consumer choice become a distraction from, or an ineffectual proxy for, a coherent landscape imperative? What if it really is all about grass?

Grasslands once dominated North America, serving as matrix and sustenance for every natural process that governs the continent's function—precipitation, pollination, infiltration, predation, migration, consumption, respiration, decomposition, mineralization, you name it. Last is runoff, the continent's export. Agriculture has altered that regime, such that now more than 80 percent (Manning reports 82) of arable land is dedicated to corn, soybeans, and a handful of other commodity crops. The continent's natural functions, and humans' dependent economies, are altered accordingly. As farmers and eaters, residing in the wonderful diversity of farmstead geography and the human marketplace, it's hard to wrap our minds around that central reality. Perhaps we need to try. I'm haunted by this continental shift in land use, for I was a geologist before I became a farmer. Nearly thirty years ago I watched as my first geology professor scrawled out calculations on the blackboard, deriving the chemistry of streams, then rivers, then oceans, and finally the atmosphere from the composition of rocks and dirt that feed them. These "back of the envelope" calculations became the draft of his graduate textbook, *Chemistry of the Atmosphere and Oceans.* I soon understood, as a geologist, how such derivations could constitute a primer for subsequent climate modelers and earth historians who revel in the lore of the geologic record. But as a geologist-turned-rancher, I have to recognize that such a primer serves also as a cautionary tale for farmers or whoever else collectively alters the continental exports through unabated soil mining.

To articulate a coherent vision of agriculture, we must start by grappling with the role of grass. Could this continent evolve into a mosaic of grassland ecosystems with a 21st century overprint on its pre-European origins? Could there be some assemblage of bison ecology on the semi-arid prairies; more intensive management in suitable microclimates; more vibrant, integrated crop/livestock systems, freed from the competitive burden of mono-crop subsidies? Such fantasies raise specific questions that are not currently undergoing research on a quantitative, national, or global scale. While the focus is on grass, it is no dismissal of *non*-grassland ecosystems, and *non*-grass-based food. To the contrary, restoration of a grassland economy is just the primary tool for displacing the subsidized corn/soybean/feedlot elephant in the living room. And only with it can we begin to restore the potential of *all* agricultural and wild systems. To turn those tables, a number of questions must begin to be addressed in regions throughout the continent.

- Can we shift the balance from a primarily grain-fed livestock toward a grass-based diet for meat animals without inflicting ecological harm?
- Conversely, what ecosystem services could a return to grass restore, if managed well and widely? (e.g. restoration of soil fertility, rare species, pollinators, natural carbon sequestration, etc.)
- Can grazing be managed in such a way that it can cope with both the seasonality of grazing and the relentless nature of consumer demand?
- What is the nature of the *human* cultural shift that would be required to implement a shift to a grass-based regime?
- Is such a cultural shift possible?
- What infrastructure and policy changes would have to evolve to enable such a system, and are we willing to make them?

- How can the issue of grass surface in all citizens' priorities: from eatèrs, to legislators, to voters, to landowners?

Such a grass farmers' research agenda seems an implausible ambition. Despite the eloquent praises and premium prices lavished upon many model and aspiring grass farmers in recent years, the systemic challenges driven by the corn-soybean-feedlot machine have not diminished, nor has the massive power of the corporations that have fostered it. At the same time, many entrepreneurs and researchers around the world rightly see that this mighty elephant in the living room is also arthritic, covered with warts, wobbly, obese, and unbalanced. Such fallibility inspires those who seek to restore grassland economies in preparation for the demise of corn/soybean infrastructure, and in recognition of a landscape imperative. Food systems have evolved before, and they will do so again.

In recent years, just as this prospect of grassland restoration has begun to gain traction, the elephant in the living room is shouldering an even bigger burden than our grocery bags: *ethanol.* We are now beginning to ask corn to serve not only as food, fructose, and feed, but fuel as well. It is daunting to raise the complex ethanol prospects, because the variable potential raw materials, distribution schemes, and byproduct economics preclude any simple verdicts on ethanol as a whole. However, regardless of the engineering details, the ethanol "burden," especially in its current subsidized form, could well abort any mission to soberly assess and restore a grassland economy. We know that all the cropland on earth cannot satiate our transportation appetite. Alternatively, perhaps the absurdity of asking the shaky elephant to let us starve ourselves in order to fill our gas tanks could be the wake-up call we need.

If we're fully awake, we'll see the *whole* landscape, with all the processes connecting east and west, urban and rural, wet and dry, affluent and poor, continental and marine, predators and prey. With, and perhaps only with, the coherence of such a landscape view, can we hook the power of our grocery habits into the same harness as our public health worries, our social instincts, our political maneuvers, and our affinity for wildlife. What a team that could be. It all starts with grass.

LIVING WITH WOLVES

LUBA VANGELOVA

In rural Bulgaria, the revival of one of the world's oldest breeds of livestock-guarding dog is helping to save a close canid relative—the wild wolf.

"If the dogs weren't here, the wolves would be—and how," says Tosho Georgiev, a park ranger who grew up in the Pirin mountains in Bulgaria's rugged southwest, where his parents Georgi and Slavka Georgiev graze their sheep. Pointing to a hill across the valley, he explains how a shepherd without dogs recently lost two sheep to wolves.

As he speaks, the Georgievs' latest litter of four Karakachan puppies—volleyball-size furballs in black and white—chase each other outside the sheep pen. Like all young canines, they're testing their limits and soon tumble into trouble, getting stuck inside a cauldron that proves easier to enter than to exit. They whimper, but neither their mother, sitting nearby, nor their owners come over to help. Self-reliance must begin at an early age if these puppies are to succeed in their dangerous line of work—fending off hungry wolves. The puppies will grow up to resemble fighting-weight Saint Bernards, standing two and a half feet at the shoulder, with long, dense, white fur with dark splotches (or the reverse). Thick, iron collars with rough spikes will protect their necks in case of attack.

Over the past four years, the Georgievs have been given five Karakachan dogs (two have since died) by the Bulgarian Biodiversity Preservation Society—Semperviva. Semperviva is collaborating with the Balkani Wildlife Society to simultaneously protect wolves and save an ancient dog breed, bred 5,000 years ago by the nomadic Karakachan tribe to protect their sheep.

In this corner of Europe on the Balkan peninsula, wolves still prowl forested mountain ranges that are sparsely settled by small-scale farmers. Conservationists consider this area, together with the Iberian peninsula, a cornerstone of the wolf's chances for survival in Europe.

During the first half of the twentieth century, Bulgaria's wolves numbered more than 1,000. From the 1950s to the '70s, unrestricted hunting and a war on rabies thinned their ranks. By the time the government granted protection for the species in 1985, fewer than 200 survived. The population slowly recovered, and by the early 1990s had increased to an estimated 500.

What has happened to the population since then is a point of contention. A 2001 government census counted 2,000 wolves. The Balkani Wildlife Society, which has studied the country's wolves since 1993, believes that count is flawed, contending there are fewer than 1,000. Balkani's chairperson, biologist Elena Tsingarska, says that some census-takers had a vested interest in inflating the figure: Hunters, for example, supplied the numbers for land controlled by hunting groups. She says, too, that because wolves roam large territories and census-takers didn't follow tracks into adjacent areas, it's likely some animals were counted twice.

Reliable or not, these latest figures have been used to justify a new law that requires hunters to kill predators, including wolves, they encounter while hunting other game. The government says it hopes to reduce the number of wolves to between 200 and 250, which it deems adequate. In a country where the average annual salary is about US$1,500, hunters receive 25 leva (US$12.50) and a cord of firewood per wolf kill; a National Forestry Board official says this merely covers hunting costs.

Public opinion appears to be on the government's side. In a 1997 Balkani survey, most respondents labeled wolves destructive and said their numbers should be reduced. Shifting the tide in the wolves' favor will require changing public perceptions. Prejudices die hard, especially in a country where many urban residents still have rural family ties, and wolf predation is a big concern. The key, Tsingarska says, is to find ways for farmers and livestock to coexist with wolves. "If shepherds are not happy with wolves, it influences everyone," says her collaborator, Semperviva president Sider Sedefchev.

So in 1997, Semperviva and Balkani interviewed farmers about livestock protection methods. The majority thought the most effective way was the old way: enlist wolves' nearest domesticated relatives for the job, just as shepherds traditionally had done. In these parts, that meant Karakachan dogs. The problem then became finding them.

Until the 1950s, these dogs accompanied the Karakachans and their flocks, who summered in Bulgaria's mountains and wintered along the milder Black or Aegean seas. Then, Bulgaria's Communists appropriated all privately owned livestock and forced the nomads to settle down. A few dogs were sent to state-run farms; the rest, now without work, were turned loose. Over the next two decades, rabies-motivated hunting depleted their numbers. When the Communist government fell in 1991, the few Karakachan dogs on state farms were released to fend for themselves. Fortunately, in some isolated mountain communities, shepherds here and there retained some pure-bred specimens.

In 1992, when Sedefchev was a student living just outside the capital, Sofia, he bought a Karakachan dog for a pet. While searching for a mate for it, he came to appreciate how rare the breed had become. Alarmed, Sedefchev and his brother launched a "manic" search, combing the countryside several days a week, stopping at taverns to quiz locals about Karakachan dog sightings. Vowing to save the breed, they founded Semperviva. They bought the few uncastrated adults they found and set up a breeding center in their grandparents' backyard, near Sofia. Sedefchev estimates that fewer than 400 pure-bred Karakachan dogs remained by that point, and that only a small percentage of these lived and worked with shepherds.

In 1997, Semperviva was able to begin giving out puppies (one male and one female) to shepherds like the Gorgiev family in the mountainous southwest, where wolf predation is the most serious. The puppies live among the sheep from the age of about eight weeks. They form strong bonds and will instinctively protect the sheep even at their own expense. Sedefchev tells how, at the turn of the century, a Karakachan shepherd lost his flock and dogs while wintering in Turkey. Failing to find them, he returned to his summer base in Bulgaria, some 300 miles away. Months later, his two dogs showed up, emaciated and weak, herding the entire flock of sheep.

Once the puppies are placed with shepherds, Balkani and Semperviva representatives visit every week or two, then every few months, to track the dogs' development and working skills, and record the frequency of wolf attacks. When the dogs are old enough to reproduce, the owners distribute their litters among other shepherds, often exchanging a puppy for a sheep.

The dogs' self-reliance and strong bonds with the sheep make them slightly aloof from people. Commands are rare and limited to "come," "go," and a show of direction. "The dogs figure out the rest," Sedefchev says. The dogs patrol the flock's perimeter, looking out for predators. They grab naps during the day and remain alert at night, when the shepherd is sleeping and wolves are most active. When the sheep are secured inside the pen, the dogs mark the surrounding area with urine and feces to warn away wolves and the occasional marauding bear. If they smell or hear a predator, they bark a warning. The wolves, however, know to approach from upwind—"that's the game between the wolves and dogs," Tosho says.

If barking fails to intimidate the predator, the dogs do a mock charge. If that doesn't scare it away, they try to land a shoulder blow, to show the animal they can expose its jugular and kill it if necessary. As a last resort, they will follow through on the threat. Semperviva's records testify to their effectiveness: The shepherds to whom they have distributed their first 30 dogs have yet to lose a single sheep or goat to predators.

While hoping that the Karakachan dogs will soften the shepherds' and ultimately the public's attitudes toward wolves, Balkani is also trying to alter public perceptions through the schools. Two years ago, Tsingarska and Sedefchev, a trained artist, created a 23-page children's activity book about Bulgaria's endangered large carnivores, and distributed it to 5,000 elementary-school children. The humorous puzzles, games, and comics teach children how these animals fit into the ecosystem. "We get them to think about who's the bigger hunter, wolf or man," Sedefchev says. Balkani also gives wolf-related slide presentations to junior-high students, and has produced an educational video.

Wolves in Bulgaria, as in much of Europe, have been little studied. So Balkani also began field work in 1997, studying wolf tracks and droppings to glean clues about the wolves' range and diet. Scat analysis from one pack showed that livestock comprised the largest percentage of its food. Tsingarska says this is probably connected to the sharp, poaching-related drop in deer and other wild prey numbers in the past decade.

This year Balkani plans to radio collar several wolves to gain a clearer picture of wolf-human flashpoints, and better insight into how to minimize such conflicts. "Our goal is to make a wolf conservation strategy," Tsingarska says, "and convince the government it can't allow uncontrolled killing."

Most Bulgarians have more pressing worries than animal welfare in this economically troubled country. Yet, there are reasons for optimism. The distinction between "harmful" and "useful" animal species is losing currency, Tsingarska says. And although the mindsets of adults can be difficult to change, the next generation's are still forming.

"Our hopes are in the children," Tsingarska says. And in the Karakachan puppies.

A PLEA FOR BEES

DANIEL IMHOFF

A vanguard of honeybees swarms around John Bayer's unprotected arms, head, and face as he lifts the lid on one of a half dozen hives platooned in an almond orchard outside Delhi, California. Though it's still mid-winter on this February morning, hundreds of thousands of acres of almond orchards throughout California are cloaked in leis of white blossoms, meaning that another pollination season has begun; by May, the trees will begin to drop their excess fruit. By late summer, the skies will thicken with swirling dust from industrial almond harvesters.

Bayer—medium height, stocky, with a head of bristly dark hair—squeezes a few puffs of smoke into the open box, then pries up a bee-laden wood-edged frame with a putty knife. He hesitates as crystalline nectar oozes back into the hive (or colony), and then holds up the honeycombed frame for me to see. Bare handed, he points to cells packed with clear honey, others with dark pollen, and still more with three-week-old eggs.

Though Bayer's honeybees have recently been revived from winter dormancy, the frame represents one of the most essential, industrious, and perhaps most vulnerable, cornerstones of modern agriculture—honeybee pollination. One out of every three bites of food arrives on your table compliments of an animal pollinator, a role usually fulfilled by bees but also by butterflies, beetles, birds, and even bats. We would by no means starve without the countless third-party agents that shuttle pollen between male and female flower organs, an ancient partnership that makes it possible for plants to produce seeds and fruit. In the absence of domesticated honeybees and other pollinators, however, the varieties, quantities, quality, and prices of our plant crops would be radically altered. Apples and pears, blueberries, melons and squashes, alfalfa, tomatoes and peppers, soybeans and cotton, are just a few of our crop varieties dependent on some form of pollination. And the single pollinator most relied on to maintain this essential cycle of life is the fragile yet stoic European honeybee, *Apis mellifera.*

Introduced to North America 400 years ago by the early English colonists to provide their settlements with honey, the European honeybee has since been established in nearly every farming region in North America. This tenuous dependence on a single specie has been replicated by agriculturalists all over the world. But the numbers may no longer be penciling out. Pests, pesticides, disease, and habitat loss have led to a 50 percent decline in North America's honeybee population since 1950. Experts say we're set up for a crash.

THE BEEKEEPER'S BURDEN

"Honeybees are extremely chemically sensitive," says Bayer, a third-generation beekeeper. "Pesticides can be fatal to bees, and the presence of chemicals makes them very angry. I plead with growers to at least spray at night when the bees aren't flying, or even better, to stop using chemicals altogether."

Farmers' determination to spray their crops with herbicides and insecticides is one reason that American agriculture faces an imminent crisis. The presence of pastures, woodlands, and other habitats where bees can safely forage without being poisoned by chemicals have become increasingly rare. In addition, parasitic mites, such as the *Varroa destructor,* attack bees. Beekeepers have also fought them with chemicals, and the mites have become more chemically resistant, making honeybees even more vulnerable.

That Bayer has any hives at all is a testament to his skill and devotion. As a commercial beekeeper, he rents his bee colonies to growers across the state, trucking them from farm to farm. With each year, the almond orchards of other farmers take greater precedence over Bayer's own, which grow on land his grandparents worked.

When I visited Bayer in February, he had just returned home from the arduous task of hauling 2,500 colonies to and from 18 almond farms throughout the Central Valley. This left only three decrepit-looking bee boxes to pollinate his own 40 acres. "Customers first," he sighs.

THE DEEPENING BEE DEFICIT

While honeybee populations have been declining steadily, the worldwide demand for almonds has boomed. High prices have led California farmers to devote more and more acreage to the nuts. With over 600,000 acres of almonds, the Golden State accounts for half of current worldwide production, generating over $1 billion annually. There is one small catch. Those almonds—*am-mins* in Central Valley lingo—are more dependent on honeybees than any other crop. To successfully set fruit, an almond flower must be "cross-pollinated," or fertilized with the pollen of a different variety. (We eat the fertilized embryos.) Because California's almond bloom occurs in mid-winter, rain, wind, and cold weather can all impact bloom and pollination. The sterile conditions of production orchards confound the situation as conventional farmers load the already polluted environment with a nasty brew of fatal pesticides—even during the day while chemically sensitive bees are flying. To compensate for these challenges, almond farmers supersaturate their orchards with honeybees, using two or more colonies per acre, requiring upward of 1.4 million colonies, or half of all commercial hives in the entire United States.

Compounding the problem has been the loss of nearly one-third of California hives from the fall of 2004 through the winter of 2005 to causes scientists are still trying to figure out. To meet the mounting demand for pollinators, growers have imported honeybees from as far away as Texas, Florida, the eastern seaboard, and the upper Midwest. And for the first time, U.S. borders were opened to honeybees from New Zealand and Australia. This transcontinental pollination system carries environmental costs, not least of which is the burning of fossil fuels. Hitchhiking along on the truck beds that carry the hives may be fire ants, beetles, and other alien invaders—plus disease. All can create even more problems when they arrive at their new homes.

The deepening honeybee deficit means a windfall for beekeepers with healthy colonies. In the spring of 2005, the fees that beekeepers charged farmers for renting their colonies doubled and in some cases even tripled overnight, from $48 to between $100 and $140 per hive. Those costs will surely be reflected in the price we pay for almonds. Eventually, however, a shortage of pollinators could result in lower yields—not just of almonds but also of all honeybee-dependent crops.

THE HIDDEN INDUSTRY

According to the American Beekeeping Federation, Bayer is one of just 2,000 commercial beekeepers (with 300 hives or more) managing approximately 3,000,000 colonies in the United States. "There aren't too many people who can say they derive their livelihood entirely from an insect," Bayer says proudly. "We are what I call 'the hidden industry' that makes agriculture possible. For every dollar that I'm paid for pollination services," he claims, "my clients generate between 50 and 1,000 dollars in crop revenues." The Xerces Society for Invertebrate Conservation tallies the total U.S. production of insect-pollinated crops at $US 40 billion annually (including meat and dairy products dependent on alfalfa).

In pursuit of pollination fees and high-quality honey, Bayer will drive upward of 50,000 miles by year's end, sometimes working 100 hours a week. He and his bees will embark on a crop-chasing whirlwind, moving from almonds to cherries and pears, to broccoli, cauliflower, and other brassicas, to feed fields of alfalfa, and finally on to cucumbers, squashes, and melons, before returning to wintering grounds where the bees can catch some R&R.

Like his fellow apiculturists, Bayer faces challenges that, if unmet, mean that millions more bees will die. *Varroa* and tracheal mites suck the life force out of adult or brooding bees, reducing a colony's ability to survive. Beekeepers have fought back with (increasingly toxic) pesticides, but that has ultimately produced new generations of chemically resistant "super mites" that are slowly gaining an upper hand. A second threat occurred with the arrival of "Africanized" bees in the early 1980s that had been interbred with traditional European varieties throughout much of the American South. These hybrid bees are extremely aggressive and fatally susceptible to cold. Then there is a contagious infection that decimates brooding bee larvae. To thwart that disease, doses of antibiotics are applied to colonies with increasing frequency, although this is spawning new antibiotic resistant strains. Finally, numerous macro-economic forces plague today's beekeepers: artificially high prices for U.S. sugar make it expensive to sustain hives through the winter; global competition keeps honey prices depressed; and insurers won't secure honeybee colonies against catastrophic losses. It shouldn't come as any surprise, then, that U.S. beekeepers, like small family farmers, find themselves atop agriculture's threatened list.

With the exception of one devastating year, Bayer has averted disaster. In 1980, as soon as Africanization became a known threat, he began the practice of overwintering his hives farther and farther out into cooler and less contaminated environments in the Sierras and along California's central coast. Relying on his own highly educated, if self-taught powers of observation and genetic selection, he has managed to breed gentle, productive, hardy bees that can survive cold temperatures and other pressures. At times he's even dabbled in industrial design. Among his inventions is a lightweight, articulated, 4-wheel drive forklift built from off-the-shelf parts to make moving pallets of hives easier.

If all this moving of bees from monoculture to monoculture seems incredibly resource intensive, imminently vulnerable, and even slightly insane, that's because it is. And this is not lost on John Bayer. Fifty years ago, he says, there were twice as many honeybee colonies in North America. Almond orchards did not stretch from Chico to Tehachapi, and farmers didn't pay beekeepers for pollination services. To make a living, a typical commercial beekeeper kept just one-tenth of the bees that his modern counterparts manage, and traveled about one-tenth the distance. Farm fields and orchards were bordered with natural habitats, as were roadsides and creek beds. These living borders attracted a diversity of insects, including important supplemental pollinators. (Out of the country's 4,500 native bee species, over 800 live in California. All, from the blue mason to the lemon-yellow bumblebee to the desert-ranging alkali bee, are at risk.)

THE FARMSCAPING SOLUTION

To see what such an integrated farming operation might look like, I traveled to the Capay Valley, a diversified farming region 200 miles to the northwest of Delhi. In a former life, Full Belly Farm was an almond operation. Today on 200 acres of orchards, pastures, and farm fields, it raises asparagus, carrots, peaches, other fruits and vegetables as well as cut flowers. Full Belly supplies farmers' markets, restaurants, and over 600 members of a Community Supported Agriculture operation.

Full Belly is also at the forefront of a movement known as "farmscaping," the integration of ground covers, plant strips, and even hedgerows of native plants designed to attract a diversity of insects to the farm. Organically certified for more than 20 years, Full Belly is also self-pollinating.

"Our goal on this farm is to grow insect biology in addition to our crops," explains Paul Muller, who helped establish Full Belly over two decades ago. As an example, he points to a tractor-width strip of waist-high mustard plants, now glowing vibrant yellow, that separate a field of strawberries from the stubble of last fall's broccoli crop. "We rely on a variety of different plants to help reduce the costs of our farming system. Some plants fix carbon and nitrogen. Others dish out a year-long feast of pollen and nectar sources for all kinds of insects, including pollinators."

For five years, a team of bee biologists at the University of California at Davis including Claire Kremen and Robbin Thorp conducted a study of pollinators that live in and around the wild margins of Full Belly, with a particular focus on watermelons. The work entailed the painstaking tasks of identifying, counting, and tagging pollinators to monitor their habits and numbers. Over 30 different pollinators were found to be thriving in Full Belly's chemical-free watermelon fields, enough native wild pollinators and feral honeybees to service the crop as well as the entire farm.

Could the "insect growing" strategy of Full Belly Farm realistically translate to almonds and other conventionally produced commodities—even as a backup strategy against honeybee declines? "We definitely need a back-up plan for pollination," explains Claire Kremen, now a professor of conservation biology at the University of California at Berkeley. Kremen and her fellow researchers regularly find as many as 50 native pollinators in organic farms surrounded by wild areas. At least seven wild bee species visit almonds during cold winter months, she reports. But in order to attract them, growers need to be willing to farmscape, and the more native habitat the better. Farms with no nearby oak woodlands or chaparral harbor too few natives to thrive without the services of honeybees. But farms set on a relatively smaller scale and tucked into hillsides or adjacent healthy streamside vegetation can support high numbers of pollinators, even for such demanding crops as almonds or watermelons (both grown at Full Belly Farm).

BACK TO THE FUTURE

At the end of the day, John Bayer is hopeful—farmers have to be. But he worries that the honeybee shortage will first have to get worse before enough people realize that the current system is unsustainable.

"I see honeybees as the canary in the mine shaft, and right now, our canary is very sick," he says. "My hope is the coming crisis will encourage the government and the industry to get behind essential research on the benefits of organic practices, and into better bee breeding. Both will help us take better care of our crops and the bees that pollinate them."

In the meantime, Bayer is planning to make changes on his own farm. "I'm thinking hard about my future," he says. "I could double the number of colonies I manage to take advantage of projected growth in the demand for almonds and to cover the escalating costs of insurance, fuel, and time." That's the sky's-the-limit trajectory the almond industry seems to be taking, but Bayer is reluctant to follow the crowd. "The other possibility is to move toward a smaller, diversified system like the one my grandfather followed—even using chickens to eat insect pests."

For now, Bayer's own 40-acre orchard outside Delhi remains a work in progress. He manages the trees organically and carpets the orchard floor with native wildflowers that both fertilize the soil and provide pollen and nectar. "Our future depends on our willingness to listen to Nature's subtle directions," he says. "And Nature is speaking to us right now."

MAKING ORGANIC WILD

JO ANN BAUMGARTNER

Not every organic farm has to contend with a mountain lion bounding out of a tree to seize livestock, but many organic farmers have an approach toward wild Nature that reflects such an accommodation. Months after the busy holiday season, I met with farm manager John McMullin at Embudo Valley Organics in northern New Mexico to learn about its free-range turkey operation, and to see how its practices could inform and influence national organic biodiversity guidelines. Walking on the farm's floodplain soils situated in between the Rio Grande's natural river corridor and the uplands beyond gave a feeling of wildness on the edges that supported an abundance of natural prey for predators. Over the years, the farm increased its precautions to discourage predators from an unnatural farm diet, but once in awhile they still lost a turkey to mountain lions. John told of how this predator ingeniously snuck by their Anatolian shepherd guard dog, jumped from the cottonwoods over their electric fence onto the "turkey tractor" pens, had a meal then left in reverse. He called this acceptance of predators "tithing to the wild." Embudo's practice of using protective measures to coexist with the wild, and in particular valuing a mountain lion's top-down predatory influence, is a model worth emulating.

WILDER FARM CONSCIOUSNESS EMERGING

On organic farms throughout the country, a wilder farm consciousness is emerging even as the movement struggles with its fast growth curve and often conventional-like approach to overprocessing and global marketing. Though some organic farmers, whose families never succumbed to the green revolution's sterile farm mentality, brought with them the gestalt of farming with Nature, the majority was not as fortunate or far-sighted. In an effort to wrestle with the most egregious aspects of modern agriculture, a large number of organic farmers focus on the latest biological techniques that do not require, for example, the application of 8 different pesticides on a 45-day lettuce crop—some so insidious that they cannot be washed off. A common tenet on organic farms is working with Nature, instead of pounding the life out of everything but the agricultural crops and livestock.

This wilder awareness on farms is an integration of new and old knowledge just as organic philosophy is an infusion of new and old values. Wendell Berry's moral and spiritual elements of our relationship to Nature and the land have influenced the movement as have Rachel Carson's warnings about the free-for-all pesticide use and a silent spring and J. I. Rodale's belief that healthy, alive soils yield healthy people. Yet even their ideas were not new. Long before them and the advent of the dominant industrial agricultural paradigm, farmers and researchers were laying the groundwork for managing food production within diverse natural systems.

In the 1890s, a precursor agency to the U.S. Department of Agriculture was formed called the Division of Economic Ornithology. These early researchers, committed to discovering which birds were most important on the farm, examined the contents of more than 80,000 bird stomachs. Thanks, that is, to farmers responding to their requests to shoot and mail in bird specimens. Those old records show an amazing amount of today's common pest insects, rodents, and weed seeds eaten: robins eating caterpillars; woodpeckers—codling moths; vireos—scale; grosbeaks—Colorado potato beetles; hawks—gophers, rabbits, mice; and sparrows—weed seeds. Back then, it was calculated that bird predation of destructive pests was worth in today's world over $14 billion. Remember, this was at a time when farms were smaller, with native pastures and diverse crops, and lacked technology to drain, scrape, and level fencerow to fencerow.

Today, the Organic Farming Research Foundation estimates that research is being conducted on about 2,000 acres of U.S. certified organic ground. Most organic research projects study crops and livestock systems. Very few examine how farm practices can support the wild. Researchers, though, are interested in biodiversity when it comes to comparing differences on organic versus conventional farms.

European analysis of 66 studies found that organic farms support more plant, bird, and predatory insect diversity and abundance than conventional farms, especially in intensively managed agricultural landscapes. Smaller and more diverse agricultural and natural landscapes did not show as many biodiversity variations between organic and conventional management. Researchers in the United States report that organic is often as important, if not more so, for healthy populations of native insect pollinators than lower-intensity farming because these species are so hard hit by pesticides. Still others report that organic orchards support greater bird fledgling success than conventional orchards with toxic pesticide use. Thus, biodiversity is positively influenced by organic methods as well as the intensity of farm scale. This research reinforces the importance of biodiversity, an aspect that has turned out to be more central to organic farming than most realized.

PROTECTING WETLANDS, WOODLANDS AND WILDLIFE—IT'S IN THE ORGANIC RULE

Remarkably, the traditional organic philosophy to adhere to and value Nature managed to stay intact when the Organic Food Production Act and the national organic standards were written. By the time it was enacted in 2002, there were industrial-scale influences competing with grassroots interests for the very future of organic agriculture, albeit interests that had been strengthening over the previous 40 years. Not only does the actual definition of organic production include promoting ecological balance and *conserving biodiversity,* the preamble to the standards explains the intent of this wording, saying "The use of 'conserve' establishes that the producer must *initiate* practices to support biodiversity and *avoid,* to the extent practicable, any activities that would diminish it." Even more important, the Organic Production and Handling Standard (205.200 General) requires that production practices "must *maintain or improve* the natural resources (soil and water quality, *wetlands, woodlands* and *wildlife*) of the operation."

A few years after the Act became federal law, the Wild Farm Alliance (WFA) was approached by the Independent Organic Inspectors Association, the group that trains about half the organic inspectors, to see if we might help organic certification agencies and inspectors understand and inspect for this regulation. WFA organized a technical advisory committee of farmers, certifiers, and conservationists to help create a set of guides for the organic community. As with many collaborations, we got side-tracked into something unexpectedly valuable. It was pointed out that the National Organic Standards Board (NOSB), the guiding body for the USDA's organic rule, would likely embrace the concepts we were identifying, if we put them into a set of biodiversity conservation inspection questions that could be used by certification agencies when inspecting and approving organic farm operations. After numerous presentations and broad-based input, the NOSB unanimously amended their model inspection forms to address biodiversity.

The term *biodiversity* (biological diversity), originally coined by conservationists, is all encompassing. It includes the variety in every life form, from bacteria and fungi to grasses, trees, and wildlife. Biodiversity also includes genetics, species, and natural communities found in an area, as well as the full range of natural processes upon which life depends, such as nutrient cycling, carbon fixation, and predation.

While the organic community as a whole is beginning to realize that biodiversity is part of the philosophy and is mandated in the USDA's organic rule, the actual meaning of biodiversity has often been interpreted vaguely and is perhaps misunderstood. Conserving biodiversity requires safeguarding soil biology and rare seed varieties and animal breeds (farm biodiversity), but it also requires protecting the native plants and animals (wild biodiversity) existing on and flowing through the farm. The sterile, monoculture reality of industrialized agribusiness has clearly shown the need to conserve both our farm heritage and the native species with the necessary functions they provide in our agricultural landscapes. Reconciling the meaning of biodiversity that has all along incorporated both the farm and the wild only makes sense. Further, the federal organic rule specifically points to conserving the wild—wetlands, woodlands, and wildlife.

Are organic certifiers taking the NOSB's guidance to heart and requiring their farmers to conserve biodiversity? Many are, but it is a slow process, and certification is a competitive business. Some leading organizations have readily adopted biodiversity inspection questions. Some intend to once they have staff time, while others continue to sit on the fence, until they receive encouragement. It may well be also that these certifiers are sideline-sitting because this is so new; they are just beginning to learn about biodiversity conservation of Nature, and there are gray areas remaining in the rule. Ambiguities existed in the interpretation of organically grown seed before the kinks were taken out, and the same is true currently for organic pasturing of animals and for biodiversity conservation. The overall concepts are in place, but the nuances must be worked out in the coming years. Organic farming *is* beginning to live up to this more expansive reputation, and as more certifiers embrace biodiversity conservation principles, a new cycle will begin in force. Consumers' purchases will be supporting wild Nature.

CONSERVING AND RESTORING BIODIVERSITY

Remnants of the wild can even still be found in California's Central Valley's industrialized agricultural landscape, especially along the river courses. John Teixeira says he is lucky to farm with wildlife habitat existing along the edges of his land between Lone Willow Slough and the San Joaquin River. His organically certified, sweet-tasting heirloom tomatoes are blessed with native diversity. Bumblebees pollinate flowers. Predatory wasps patrol for aphids. Dragonflies munch on mosquitoes and feast on flies throughout the hot summer. The nearby riparian habitat, with something usually in bloom—the cow parsnip, wild sunflowers, rosehips, stinging nettles, willows, cottonwoods, and many other plants—provides floral food sources and shelter all year. As evening rolls around, bats turn up scouring for flying insects. Later the night becomes alive with great horned owls hooting and coyotes singing in the distance. John has happily seen a pack of five mousing. These predators, along with skunks and foxes, move through vegetated corridors on their nighttime routes doing what they've done for eons. Only in this case, they are sharing their hunting skills in exchange for their *pièce de resistance*: organically raised farm pests. And, of course, safe passage in an often toxic and persecuting world.

Organic farmers have a head start on most of their conventional counterparts in that they have already been encouraging some aspect of Nature, trying to understand and mimic the complex environment. It is not always the case, though, in areas where the green revolution has obliterated all but the last vestiges of wild Nature. Understanding the intimate connections of what was there previously is not all that obvious—much less imagining its beauty and functionality—when thousands of acres of a few monocrops in a region surround the organic farm as far as the eye can see. It is possible to bring back some of the wild processes even in these areas as Tom and Denesse Willey have done on their 80 acre organic, diverse vegetable farm. Just 20 miles from John Teixeira's, but a world away and some 13,000 conventional monocrop acres in between, they have established a native plant hedgerow. Their crops are now receiving natural pollination services from wild bees that amazingly find this toxic-free habitat in the otherwise ecologically barren setting of the region. When allowed and supported, even at the smallest scale, Nature moves in. With the respect the Willeys have earned in their community, other farmers may be following their lead and creating similar habitat as well.

Biodiversity concepts are resonating on organic farms, for the most part, because they are just common sense. When farmers realize that habitat loss is the greatest threat to native plants and wildlife, the importance of heightened stewardship becomes clear. Depending on farmers' goals they may start by conserving what they have and continue by restoring habitat, putting small clumps of native plants in unused areas under telephone poles or next to pumps, or possibly installing larger native plant hedgerows along roads to attract beneficial insects, as the Willeys have. Ranchers may start by bonding protective llamas or dogs to their livestock in order to scare away all but the largest predators; may choose to implement rotational grazing where intensive management results in higher quantity and quality of diverse forage, fewer weeds, and increased biodiversity; or may closely monitor grazing near riparian areas and fence livestock out of these important ecosystems, if daily monitoring is not possible.

When habitat is protected from overgrazing and, if need be, restored with native plants adjacent to ponds, creeks, and streams, it helps to filter and clean water and recharge groundwater. In fact, good water quality, a universal value in urban and rural areas, is becoming more regulated in agriculture. So beyond the value to the farm, there are obvious conservation values in providing wildlife habitat and clean water.

BROAD-REACHING EFFORTS

Before final adoption of biodiversity measures by the NOSB, many of these organic farmers helped us test and streamline the inspection questions. On a continuum of conservation practices, a number were taking small but important steps that might seem inconsequential to conservationists, while others implemented broader-reaching efforts, such as providing habitat when their farms are on migratory routes and movement corridors, making sure there is dependable passage for wildlife. Farmers invited researchers and naturalists to help build up rare fish populations, identified sensitive habitats that should not be grazed, restored vegetation for riparian-dependant species, or actually planted crops for rare birds.

We met fourth-generation farmer Kevin Lunny on a foggy afternoon at his ranch north of San Francisco, which is now part of the Point Reyes National Seashore. A certified organic farmer, Kevin fences his cattle away from dune grasses that support threatened snowy plovers and is installing solar and gravity-fed watering stations so that the livestock do not impact rare red-legged frog habitat. Beyond that, he restored his annually tilled hay fields to native grass pastures in order to market certified grass-fed beef and better support the diverse wildlife present. During the transition, he used ultrasound to check the live tissue of different cattle breeds that had not lost their inherent ability to graze, and came up with a more efficient and tasty product. The cattle now feed themselves much of the year, instead of being fed silage. And Kevin is on the tractor less, and his soils are protected year round from

Core principles

- Take advantage of Nature's ecosystem services: pollination, pest control, beneficial predation, advantageous fire, flood and erosion control, nutrient cycling, and improved water quality and quantity.
- Avoid conversion of sensitive habitats to agricultural production or development.
- Protect threatened and endangered species, species of special concern, and keystone species.
- Conserve and restore native plants and animals of the production operation, including in and around water bodies.
- Conduct restoration based on native species and ecosystems present on the land before it was turned over to agriculture.
- Maintain and restore linkages and connectivity including large blocks of habitat and wildlife corridors to strengthen regional networks of conservation areas.
- Prevent introduction and spread of non-native, invasive species.

wind and water erosion. This change means his ranch has a more diverse population of protected species, a dozen in all, than any of the other operations in the area. He is not only streamlining the ranch to gain from Nature's principles, but also restoring grassland functions and, with them, creating opportunities for rarer species.

The give-and-take that occurs with native plants and wildlife on many organic farms is as much an art as a science. On a cool spring morning before full-on production hit No Cattle Farm in New Mexico, we met with Sharlene Grunerud and Michael Alexander to learn about these interactions. Years earlier, ash-throated flycatchers, birds that "hawk" insects in midair, were hunting over their vegetable crops when they decided to put up rebar perches throughout the fields to encourage more efficient hunting, but they did not stop there. Knowing that birds are territorial, and wanting to ensure that they come back every year, several bird nest boxes were installed away from the communal "hunting" grounds. Beyond the fields, we walked through thickets of wild grape, currant, hackberry, blackberry, black walnut, and box elder, which they value for the food, cover, and resting stops for many avian pest-eating species. We

Conservation Planning on an Organic Farm

- **Inventory your farm** for wildlife and dominant native plants.
- **Obtain a topographical map** of the watershed that shows your farm and nearby public natural resource lands and protected areas.
- **Create a farm map** that includes hedgerows, woodlands, wetlands, waterways and riparian zones, hydrological and drainage conditions, wildlife corridors, invasive species, perennial cover, topography, soils, eroded areas, and special habitats like those used by priority species.
- **Research what species lived on the land prior to farming** and locate the closest intact ecosystems that resemble the land's original state.
- **Assess the farm for opportunities to support priority species and habitats** in your watershed. These include threatened and endangered species, species of concern, and keystone species; migration and movement of native species; and ecosystem processes.
- **Find out about biodiversity conservation actions being taken by neighboring farmers and organizations** in the area and how you might contribute to or become a leader in a biodiversity strategy for the landscape.
- **Investigate incentive programs** to assist with planning and implementation, such as habitat conservation programs through state or federal agencies or through non-governmental organizations.
- **Prioritize actions to conserve biodiversity** based on consideration of regional conservation goals, priority species and sensitive habitats, invasive species and eroded areas, the conservation activities of other land managers in the watershed, and value to the farming operation.
- **Create a conservation component of the organic farm plan** with a clear definition of goals, expectations, and a timeline for implementation of conservation practices, factoring in monitoring and plan revision.

continued past large brush piles of orchard prunings and felled invasive Siberian elm trees that were intentionally spaced out evenly across the farm for Gambel's quail and sparrow habitat. Their hundred-foot cottonwoods with thick willow understory along the Mimbres River were habitat for rare Mexican black hawks, kingfishers, blue herons, snipes and ravens—one of which they were delighted to see snatch a gopher from its hole. Sharlene and Michael value how important wildness is to their farm and their psyche; they told of seeing bull elk from their bedroom door, a bobcat pushing his nose against the glass, coyotes frolicking, and a fox watching them as they watched it walking around their house.

Amish farmer David Kline writes eloquently of the daily interactions he has with wildlife. He takes pleasure in watching water pipits and pectoral sandpipers feeding on the freshly turned earth abounding with life while he is plowing his fields, and eagerly awaits the quacking and gabbling of the first of thousands of waterfowl arriving at his safe farm pond. His observations have led him to leave at least one field of mixed hay for his livestock and for birds like bobolinks that are in decline. Most of Ohio's prairie habitat went under the plow to corn and soybeans, or to pure stands of alfalfa which he terms "biological deserts except for a few biting insects." By waiting to cut this mixed hay in the latter part of June, he reports that in most cases the bobolinks have safely fledged their first broods. His is a multifaceted, ecological, passionate study. Even his days off are spent with his family getting to know the wilder areas of the farm better.

WHAT THE FUTURE HOLDS

Finances keep getting tighter on the farm and that translates into less time to study these interactions, less time for family and neighbors, and often no time to enjoy the natural aspects of the farm. The constant growth model of industrial agriculture with its market influences puts small and large farmers in opposition to the principles of Nature. Exploitation and destruction of resources has been what much of modern agriculture is about—mining the fertility of the soil and destroying habitat, all with a tremendous dependence on external energy. Aldo Leopold wrote long ago about the need of a conservation ethic to transform our society and economy, and one could argue that such an awareness is beginning to emerge both in conventional and organic agriculture. But it is just a beginning and many key challenges remain.

How will organic farming change in the coming years as energy prices increase and more big buyers like Wal-Mart get into globally trading organic products? Will sensitive lands with rare species or ecosystems be converted to agriculture in a rush to fulfill both growing markets and the three-year no-pesticide application requirement? Or will organic farmers spend more time planting native habitat that provides resiliency against pest outbreaks, soil losses, and weedy invasives? Can we trust that organic farmers will embrace a holistic approach that not only supports self-sustaining beneficial organisms, but also self-sustaining native species and ecosystems for their own sake? These questions point to the many unknowns that exist around this new and logical extension of organic farming. Yet one thing is certain. Food is that common denominator we all share, a deep daily connection with the earth. And it is hard to imagine a more joyous and profound connection than this one between organic farming and wild Nature.

Section III

BIODIVERSITY CHALLENGE

Aldo Leopold wrote that the great discovery of the 20th century was not the television or radio but the "complexity of the land organism." (Unfortunately, this revelation remains highly underappreciated.) To Leopold and the legions of ecologists, farmers, and others who followed him, protecting the whole of the land's complexity is implicit to the discipline of conservation. "Harmony with the land," wrote Leopold in *Round River,* "is like harmony with a friend; you cannot cherish his right hand and chop off his left. That is to say, you cannot love game and hate predators; you cannot conserve the waters and waste the ranges; you cannot build the forest and mine the farm. The land is one organism."

Maintaining the wholeness or integrity of the land organism may be a difficult concept to grasp, particularly on lands that have been altered and fragmented or within economic and political systems that place no value on it. Yet this is the task at hand. For agriculture to become truly sustainable, we must stand tall in the face of current pressures to reduce conservation to an exercise in resource management. The tendency to gauge success by sheer numbers of a given species or to weight all species equally must give way to a nuanced understanding of the actual conditions of biodiversity on a landscape. The important balancing functions of predator-prey relationships must ultimately be restored, even if we experience uneasiness with animals that are more powerful and larger than we are. As many will attest, there is much to be gained from learning about and garnering life-long acquaintances with the wild members of farms and nearby intact natural areas. The knowledge and historical documentation that rural lands are more than just pretty—that they function as core habitat for a multitude of species—can translate into positive conservation outcomes. In the long term, planning for durability and resilience of the land organism may ensure that Nature survives our inevitable tinkering.

A BIOTIC VIEW OF LAND

ALDO LEOPOLD

In pioneering times wild plants and animals were tolerated, ignored, or fought, the attitude depending on the utility of the species.

Conservation introduced the idea that the more useful wild species could be managed as crops, but the less useful ones were ignored and the predaceous ones fought, just as in pioneering days. Conservation lowered the threshold of toleration for wildlife, but utility was still the criterion of policy, and utility attached to species rather than to any collective total of wild things. Species were known to compete with each other and to cooperate with each other, but the cooperations and competitions were regarded as separate and distinct; utility as susceptible of quantitative evaluation by research. For proof of this we need look no further than the bony framework of any campus or capitol: department of economic entomology, division of economic mammalogy, chief of food habits research, professor of economic ornithology. These agencies were set up to tell us whether the red-tailed hawk, the gray gopher, the lady beetle, and the meadowlark are useful, harmless, or injurious to man.

Ecology is a new fusion point for all the natural sciences. It has been built up partly by ecologists, but partly also by the collective efforts of the men charged with the economic evaluation of species. The emergence of ecology has placed the economic biologist in a peculiar dilemma: with one hand he points out the accumulated findings of his search for utility, or lack of utility, in this or that species; with the other he lifts the veil from a biota so complex, so conditioned by interwoven cooperations and competitions, that no man can say where utility begins or ends. No species can be "rated" without the tongue in the cheek; the old categories of "useful" and "harmful" have validity only as conditioned by time, place, and circumstance. The only sure conclusion is that the biota as a whole is useful, and biota includes not only plants and animals, but soils and waters as well.

In short, economic biology assumed that the biotic function and economic utility of a species was partly known and the rest could shortly be found out. That assumption no longer holds good; the process of finding out added new questions faster than new answers. The function of species is largely inscrutable, and may remain so.

When the human mind deals with any concept too large to be easily visualized, it substitutes some familiar object which seems to have similar properties. The "balance of nature" is a mental image for land and life which grew up before and during the transition to ecological thought. It is commonly employed in describing the biota to laymen, but ecologists among each other accept it only with reservations, and its acceptance by laymen seems to depend more on convenience than on conviction. Thus "nature lovers" accept it, but sportsmen and farmers are skeptical ("the balance was upset long ago; the only way to restore it is to give the country back to the Indians"). There is more than a suspicion that the dispute over predation determines these attitudes, rather than vice versa.

To the lay mind, balance of nature probably conveys an actual image of the familiar weighing scale. There may even be danger that the layman imputes to the biota properties which exist only on the grocer's counter.

To the ecological mind, balance of nature has merits and also defects. Its merits are that it conceives of a collective total, that it imputes some utility to all species, and that it implies oscillations when balance is disturbed. Its defects are that there is only one point at which balance occurs, and that balance is normally static.

If we must use a mental image for land instead of thinking about it directly, why not employ the image commonly used in ecology, namely the biotic pyramid? With certain additions hereinafter developed it presents a truer picture of the biota. With a truer picture of the biota, the scientist might take his tongue out of his cheek, the layman might be less insistent on utility as a prerequisite for conservation, more hospitable to the "useless" cohabitants of the earth, more tolerant of values over and above profit, food, sport, or tourist-bait. Moreover, we might get better advice from economists and philosophers if we gave them a truer picture of the biotic mechanism.

I will first sketch the pyramid as a symbol of land, and later develop some of its implications in terms of land use.

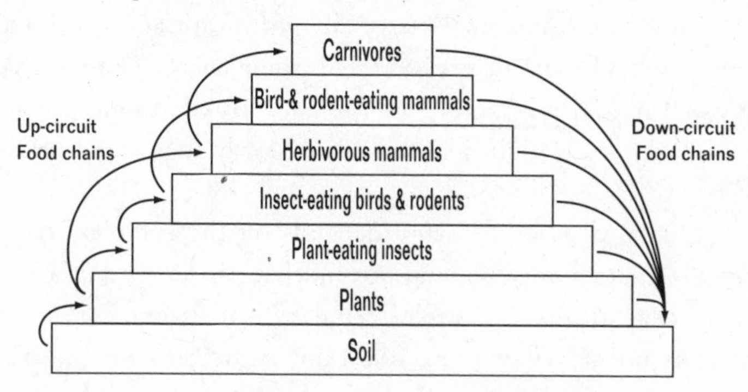

Plants absorb energy from the sun. This energy flows through a circuit called the biota. It may be represented by the layers of a pyramid. The bottom layer is the soil. A plant layer rests on the soil, an insect layer on the plants, and so on up through various groups of fish, reptiles, birds, and mammals. At the top are predators.

The species of a layer are alike not in where they came from, nor in what they look like, but rather in what they eat. Each successive layer depends on those below for food and often for other services, and each in turn furnishes food and services to those above. Each successive layer decreases in abundance; for every predator there are hundreds of his prey, thousands of their prey, millions of insects, uncountable plants.

The lines of dependency for food and other services are called food chains. Each species, including ourselves, is a link in many food chains. Thus the bobwhite quail eats a thousand kinds of plants and animals, *i.e.*, he is a link in a thousand chains. The pyramid is a tangle of chains so complex as to seem disorderly, but when carefully examined the tangle is seen to be a highly organized structure. Its functioning depends on the cooperation and competition of all its diverse links.

In the beginning, the pyramid of life was low and squat; the food chains short and simple. Evolution has added layer after layer, link after link. Man is one of thousands of accretions to the height and complexity of the pyramid. Science has given us many doubts, but it has given us at least one certainty; the trend of evolution is to elaborate the biota.

Land, then, is not merely soil; it is a fountain of energy flowing through a circuit of soils, plants, and animals. Food chains are the living channels which conduct energy upward; death and decay return it to the soil. The circuit is not closed; some energy is dissipated in decay, some is added by absorption, some is stored in soils, peats, and forests, but it is a sustained circuit, like a slowly augmented revolving fund of life.

The upward flow of energy depends on the complex structure of the plant and animal community, much as the upward flow of sap in a tree depends on its complex cellular organization. Without this complexity normal circulation would not occur. Structure means the characteristic numbers, as well as the characteristic kinds and functions of the species.

This interdependence between the complex structure of land and its smooth functioning as an energy circuit is one of its basic attributes.

When a change occurs in one part of the circuit, many other parts must adjust themselves to it. Change does not necessarily obstruct the flow of energy; evolution is a long series of self-induced changes, the net result of which has been probably to accelerate the flow; certainly to lengthen the circuit.

Evolutionary changes, however, are usually slow and local. Man's invention of tools has enabled him to make changes of unprecedented violence, rapidity, and scope.

One change is in the composition of floras and faunas. The larger predators are lopped off the cap of the pyramid; food chains, for the first time in history, are made shorter rather than longer. Domesticated species are substituted for wild ones, and wild ones moved to new habitats. In this world-wide pooling of faunas and floras, some species get out of bounds as pests and diseases, others are extinguished. Such effects are seldom intended or foreseen; they represent unpredicted and often untraceable readjustments in the structure. Agriculture science is largely a race between the emergence of new pests and the emergence of new techniques for their control.

Another change affects the flow of energy through plants and animals, and its return to the soil. Fertility is the ability of soil to receive, store, and return energy. Agriculture, by overdrafts on the soil, or by too radical a substitution of domestic for native species in the superstructure, may clog the channels of flow or deplete storage. Soils depleted of their stores wash away faster than they form. This is erosion.

Waters, like soils, are part of the energy circuit. Industry, by polluting waters, excludes the plants and animals necessary to keep energy in circulation.

Transportation brings about another basic change: the plants or animals grown in one region are consumed and return to the soil in another. Thus the formerly localized and self-contained circuits are pooled on a world-wide scale.

The process of altering the pyramid for human occupation releases stored energy, and this often gives rise, during the pioneering period, to a deceptive exuberance of plant and animal life, both wild and tame. These releases of biotic capital tend to becloud or delay the penalties of violence.

This thumbnail sketch of land as an energy circuit conveys three ideas more or less lacking from the balance of nature concept:

1. That land is not merely soil
2. That the native plants and animals kept the energy circuit open; others may or may not
3. That man-made changes are of a different order than evolutionary changes, and have effects more comprehensive than is intended or foreseen.

These ideas, collectively, raise two basic issues: Can the land adjust itself to the new order? Can violence be reduced?

Biotas seem to differ in their capacity to sustain violence. Western Europe, for example, carries a far different pyramid than Caesar found there. Some large animals are lost; many new plants and animals are introduced, some of which escape as pests; the remaining natives are greatly changed in distribution and abundance. Yet the soil is still fertile, the waters flow normally, the new structure seems to function and to persist. There is no visible stoppage of the circuit.

Western Europe, then, has a resistant biota. Its processes are tough, elastic, resistant to strain. No matter how violent the alterations, the pyramid, so far, has developed some new *modus vivendi* which preserves its habitability for man and for most of the other natives.

The semiarid parts of both Asia and America display a different reaction. In many spots there is no longer any soil fit to support a complex pyramid, or to absorb the energy returning from such as remains. A cumulative process of wastage has set in. This wastage in the biotic organism is similar to disease in an animal, except that it does not culminate in absolute death. The organism recovers, but at a low level of complexity and human habitability. We attempt to offset the wastage by reclamation, but where the regimen of soils and waters is disturbed it is only too evident that the prospective longevity of reclamation projects is short.

The combined evidence of history and ecology seems to support one general deduction: the less violent the man-made changes, the greater the probability of successful readjustment in the pyramid. Violence, in turn, would seem to vary with human population density; a dense population requires a more violent conversion of land. In this respect, America has a better chance for nonviolent human dominance than Europe.

It is worth noting that this deduction runs counter to pioneering philosophy, which assumes that because a small increase in density enriched human life, that an indefinite increase will enrich it indefinitely. Ecology knows of no density relationship which holds within wide limits, and sociology seems to be finding evidence that this one is subject to a law of diminishing returns.

Whatever may be the equation for men and land, it is improbable that we as yet know all its terms. The recent discoveries in mineral and vitamin nutrition reveal unsuspected dependencies in the up-circuit; incredibly minute quantities of certain substances determine the value of soils to plants, of plants to animals. What of the down-circuit? What of the vanishing species, the preservation of which we now regard as an aesthetic luxury? They helped build the soil; in what unsuspected ways may they be essential to its maintenance? Professor Weaver proposes that we use prairie flowers to reflocculate the wasting soils of the dust bowl; who knows for what purpose cranes and condors, otters and grizzlies may some day be used?

Can the violence be reduced? I think that it can be, and that most of the present dissensions among conservationists may be regarded as the first gropings toward a nonviolent land use.

For example, the fight over predator control is no mere conflict of interest between field-glass hunters and gun-hunters. It is a fight between those who see utility and beauty in the biota as a whole, and those who see utility and beauty only in pheasants or trout. It grows clearer year by year that violent reductions in raptorial and carnivorous species as a means of raising game and fish are necessary only where highly artificial (i.e., violent) methods of management are used. Wild-raised game does not require hawkless coverts, and the biotically educated sportsman gets no pleasure from them.

Forestry is a turmoil of naturalistic movements.

Thus, the Germans, who taught the world to plant trees like cabbages, have scrapped their own teachings and gone back to mixed woods of native species, selectively cut and naturally reproduced (*Dauerwald*). The "cabbage brand" of silviculture, at first seemingly profitable, was found by experience to carry unforeseen biotic penalties: insect epidemics, soil sickness, declining yields, foodless deer, impoverished flora, distorted bird population. In their new Dauerwald the hard-headed Germans are now propagating owls, woodpeckers, titmice, goshawks, and other useless wildlife.

In America, the protests against radical "timber stand improvement" by the C.C.C. and against the purging of beech, white cedar, and tamarack from silvicultural plants are on all fours with Dauerwald as a return to nonviolent forestry. So is the growing skepticism about the ultimate utility of exotic plantations. So is the growing alarm about the epidemic of new Kaibabs, the growing realization that only wolves and lions can insure the forest against destruction by deer and insure the deer against self-destruction.

We have a whole group of discontents about the sacrifice of rare species: condors and grizzlies, prairie flora and bog flora. These, on their face, are protests against biotic violence. Some have gone beyond the protest stage: witness the Audubon researches for methods of restoring the ivory-billed woodpecker and the desert bighorn; the researches at Vassar and Wisconsin for methods of managing wildflowers.

The wilderness movement, the Ecological Society's campaign for natural areas, the German *Naturschutz,* and the international committees for wildlife protection all seek to preserve samples of original biota as standards against which to measure the effects of violence.

Agriculture, the most important land use, shows the least evidence of discontent with pioneering concepts. Conservation, among agricultural thinkers, still means conservation of the soil, rather than of the biota including the soil. The farmer must by the nature of his operations modify the biota more radically than the forester or the wildlife manager; he must change the ratios in the pyramid and exclude the larger predators and herbivores. This much difference is unavoidable. Nevertheless it remains true that the exclusions are always more radical than necessary; that the substitution of tame for wild plants and the annual renewal of the plant succession creates a rich habitat for wildlife which has never been consciously utilized except for game management and forestry. Modern "clean farming," despite its name, sends a large portion of its energy into wild plants; a glance at the aftermath of any stubble will prove this. But the animal pyramid is so simplified that this energy is not carried upward; it either spills back directly into the soil, or at best passes through insects, rodents, and small birds. The recent evidence that rodents increase on abused soils (animal weed theory) shows, I think, a simple dearth of higher animal layers, an unnatural downward deflection of the energy circuit at the rodent layer. Biotic farming (if I may coin such a term) would consciously carry this energy to higher levels before returning it to the soil. To this end it would employ all native wild species not actually incompatible with tame ones. These species would include not merely game, but rather the largest possible diversity of flora and fauna.

Biotic farming, in short, would include wild plants and animals with tame ones as expressions of fertility. To accomplish such a revolution in the landscape, there must of course be a corresponding revolution in the landholder. The farmer who now seeks merely to preserve the soil must take account of the superstructure as well; a good farm must be one where the wild fauna and flora has lost acreage without losing its existence.

It is easy, of course, to wish for better kinds of conservation, but what good does it do when on private lands we have very little of any kind? This is the basic puzzle for which I have no solution.

It seems possible, though, that prevailing failure of economic self-interests as a motive for better private land use has some connection with the failure of the social and natural sciences to agree with each other, and with the landholder, on a common concept of land. This may not be it, but ecology, as the fusion point of sciences and all the land uses, seems to me the place to look.

NATURE'S CRISIS

DAVE FOREMAN

In my 35 years as a conservationist, I have never beheld such a bleak and depressing situation as I see today. The evidence for my despair falls into three categories: the state of Nature, the power of anticonservationists, and appeasement and weakness within the conservation and environmental movements. I fear that on some level we must recognize that this state of affairs may be inevitable and impossible to turn around. That is the coward's way out, though. The bleakness we face is all the more reason to stand tall for our values and to not flinch in the good fight. It is important for us to understand the parts and pieces of our predicament, so we might find ways to do better.

THE STATE OF NATURE

I've recently written a book, *Rewilding North America,* that goes into considerable detail describing and trying to understand the Seven Ecological Wounds that drive the Sixth Great Extinction, which is the fundamental fact and problem in the world today. Around the world, direct killing of wildlife, habitat destruction, habitat fragmentation, loss of ecological processes, invasion by exotic species and diseases, ecosystem pollution, and catastrophic climate change are worsening. We six-and-a-half-billion too-clever apes are solely to blame. Despite impressive successes here and there, the overall state of Nature continues to decline. This is simple reality, despite the scolding we hear not to be doom-and-gloomers.

POWER OF THE ANTICONSERVATIONISTS

In the United States, the federal government has become the sworn enemy of conservation. Not only has the radical-right presidency and Congress stopped any progress in the conservation and restoration of Nature, they are dedicated to overthrowing the twentieth century's legacy of conservation and environmental policy and programs. They are unabashedly trying to go back to the unfettered, uncaring era of the robber barons in the late nineteenth century. This revolution is both philosophical and practical. Bad as this is, the radical-right is also dedicated to shredding science, particularly biology, and time-traveling back to before the Enlightenment.

While the United States is an extraordinary political case, elsewhere some of the supposedly most civilized nations on the planet, such as Canada, Norway, and Japan, are again waging nineteenth-century crusades against wild Nature: frontier-forest mining, slaughter of troublesome animals (such as seals, wolves, bears), and commercial whaling, just for starters. Japanese, European, Chinese, and American businesses are looting the last wild places for timber, pulp, wildlife, minerals, and oil, opening up such places to further habitat destruction and bushmeat hunting by local people.

Although the radical-right control of the U.S. presidency and Congress was gained by a very small margin in 2004 (no mandate), it is backed by powerful and popular forces and by a shocking descent into prescientific irrationality by large sections of the public.

APPEASEMENT AND WEAKNESS IN THE CONSERVATION AND ENVIRONMENTAL MOVEMENTS

The efforts to protect wild Nature and to clean up pollution face internal subversion from the right and left that leads to deep compromises not only on issues but also on fundamental principles. We can stuff these calls to compromise into several boxes, including sustainable development, resourcism, Nature deconstruction, politically correct progressivism, and anthropocentric environmentalism.

First, some brief definitions: *conservation* is the movement to protect and restore wildlands and wildlife (Nature for its own sake); *resourcism* or *resource conservation* is the resource extraction ideology of the U.S. Forest Service and other agencies (multiple-use/sustained yield); *environmentalism* is the campaign to clean up pollution for human health and make cities livable.

The radical right has been disciplined about thinking and acting for the long term; we have failed in part because we do not have a long-term strategy to which we stick.

Internationally since the 1980s, conservation efforts to protect wildlands and habitat by means of national parks, game reserves, and other protected areas have been severely compromised as financial-aid agencies and even some top international conservation groups have shifted to promoting so-called sustainable development and community-based conservation. Although these approaches are sometimes sound conservation tactics, in practice they have elbowed Nature into second place. This establishment undercutting of Nature conservation has been joined by the leftist passion of some anthropologists and other social engineers to reject protected areas in favor of indigenous extractive reserves. Shockingly, sustainable development is coming close to dominating the pages even of publications about conservation biology and is gaining more and more adherents in resource management graduate schools and large "conservation" organizations. Some members of the academic left have become deconstructors of Nature, denying that it independently exists, proclaiming that we invent it; therefore there is no reason to protect it.

Pressured from the left and right during the last twenty-five years, conservation and environmental organizations worldwide have moved away from forthright calls for zero population growth, even though human overpopulation is the underlying cause of all conservation and environmental problems. We hear a growing drumbeat that there is a dearth of births and that developed nations face economic collapse because of fewer young people. We are essentially silent in response to this cornucopian madness. Similarly, the conservation and environmental movements in general shy away from acknowledging the reality of human-caused mass extinction. If we don't even clearly state the problem, how can we do anything about it?

We can also see a shift in the United States from conservation to resourcism among several prominent and influential entities. Once the preeminent conserver of biological diversity, The Nature Conservancy has been steadily moving to a resourcist approach. They talk now of "working landscapes," a fancy euphemism for logging and livestock grazing, and demand that their employees talk about people instead of Nature. *High County News,* once a feisty voice for grassroots conservationists in the West, has steadily turned into a voice for resourcism—not the preservation of wilderness, but the preservation of happy little resource-extraction communities—and for negotiated settlements between conservationists and resource-extraction industries, which usually favor industry.

Some consultants, foundations, and political realists are urging grassroots wilderness groups to compromise in order to pass wilderness legislation that may or may not adequately protect existing wilderness. This encouragement of appeasement is based on a desire to pass bills and is an overreaction to the narrow victory of the radical right in the 2004 election. Another source for this push to compromise is the fuzzyheaded wish that if people only talk together, everything can be worked out.

Several bright young men have gained a disturbing amount of attention with their recent speeches about the "death" of environmentalism. Insofar as they consider Nature protection at all, they demand that conservationists drop their priorities to focus on social justice and other anthropocentric progressive causes. Overall, they call on environmental organizations to essentially go out of business and just become part of the progressive wing within the Democratic Party. The overwhelming identification of environmentalism with the progressive movement and the Democratic Party is a key reason that it lacks credibility with much of the American public.

Just as there has been a disturbing shift in attitudes among large segments of the American public, so have there been problematic changes among members of the conservation public. To be blunt, many of the employees and activists with conservation groups are ignorant of our history and have not read the classic books of conservation. There is an appalling lack of intellectual curiosity in the movement. On the whole, the radical right and grassroots anticonservationists both read and think more than do conservationists and environmentalists. As far as outdoor recreation goes, young people, who once would have been hikers and backpackers, now seek thrills on mountain bikes and thus cut themselves off from experiencing Nature and from having self-interest in protecting roadless areas. I don't see kids out messing around in little wild patches; they're inside, plugged in to a virtual reality.

These are trends. Of course there are exceptions. Dwelling on the exceptions, though, keeps us from doing something about the real problems. I'm not doing "nuance" here. This sober, unapologetic cataloging of the array of problems Nature conservationists face is, I am convinced, the first step in developing a more effective strategy.

In December of 1776, the American Revolution was in its darkest hour. In response, Tom Paine wrote his first "Crisis" paper:

> *These are the times that try men's souls. The summer soldier and the sunshine patriot will in this crisis, shrink from the service of his country; but he that stands it now, deserves the love and thanks of man and woman.*

General Washington had the paper read to his miserable, disheartened troops in their frozen winter camps. There was no surrender. Years of hard battle lay ahead but victory was gained.

We need Tom Paine conservationists in *our* dark hour. Let us not apologize for loving wild Nature, for caring about other species, for speaking the truth. Reach out to others. Make deals when they are good deals. But let us not be frightened and browbeaten into appeasement. Let us instead offer a bold, *hopeful* vision for how wilderness and civilization can live together.

CONTEXT MATTERS

REED NOSS

I leaned against my dusty 1974 Opel in the parking lot of Sugarcreek Nature Reserve in southwestern Ohio, reviewing the day's count. The 10 most common bird species—common grackle, northern cardinal, carolina chickadee, red-winged blackbird, American goldfinch, blue-gray gnatcatcher, European starling, indigo bunting, field sparrow, brown-headed cowbird—contrasted sharply with some of the least common: ovenbird, black-and-white warbler, cerulean warbler, yellow-throated warbler, wood thrush, scarlet tanager, pileated woodpecker. As every ornithologist familiar with this region knows, the latter species are characteristic birds of the eastern deciduous forest of North America. In an intact forest, they are common. They belong. The former group, which dominated my counts, constitutes generalists or species of the forest edge, birds that thrive in forests degraded by human activities. These species do not need the help of conservationists to survive. What might need help in Sugarcreek, it appeared, are birds of the forest interior. The only common (within the top 20) forest-interior birds in my surveys throughout the summer of 1978 were the red-eyed vireo and acadian flycatcher.

Clearly, I was dealing with a disrupted ecosystem. In today's world, that is hardly a startling discovery. What really bothered me, however, was that Sugarcreek Reserve contained one of the larger forest patches remaining in this part of Ohio, yet the reserve's managers apparently did not recognize the value of this forest. Here, in this 228-hectare reserve and the smaller woodlots still (at the time of my surveys) connected to it by wooded corridors, was an opportunity to maintain and restore a vestige of the great eastern deciduous forest. This forest once stretched from northern Florida to central Ontario, from the Atlantic coast to the Great Plains. It was magnificent.

Part of Sugarcreek had been farmed, and the abandoned fields were in the process of returning to forest. Osage-orange and honeylocust were spreading out from the fencerows where they had been planted long ago, while boxelder, white ash, wild black cherry, hickories, and other trees sprang up from the fields of grasses and goldenrods, which once had held corn. The forest was recovering. Yet the managers of Sugarcreek seemed bent on halting this process of succession. They were cutting down areas of regenerating forest and managing them as meadows or thickets. They were mowing insanely wide trails—up to 7 meters in some cases—apparently to accommodate large groups of visitors, but in the process they were fragmenting the forest and creating abundant edge habitat. What were they thinking?

Finally, it dawned on me. The managers were very concerned about *content,* in particular, how many kinds of habitats the reserve contained and the variety of experiences they offered visitors. But they ignored *context,* including the ecological history of the region and the nature of the landscape that now surrounded Sugarcreek Reserve. Sugarcreek, its tenuous connections to nearby woodlots notwithstanding, was a virtual island of mature and recovering forest in a sea of cornfields and rapidly proliferating suburbs. It was perhaps big enough to sustain, at least in the short term, populations of the forest-interior birds that dominated the region before Europeans arrived. That is, it might have been big enough, had it been managed for mature forest habitat. Managed instead for maximum habitat diversity and interspersion, it was probably inadequate for the species that needed the reserve most. Indeed, some of the characteristic forest birds that barely hung on at the time of my surveys, such as the black-and-white warbler, had disappeared when surveys were conducted again for the *Ohio Breeding Bird Atlas,* just a few years later.

In the Sugarcreek story lies a conservation paradox. Management actions undertaken at a local scale to increase diversity might have an opposite effect on a regional and ultimately global scale. This is one of the big problems with using the term "biodiversity" to mean the simple number of species in an area, without regard to their composition, relative abundances, or habitat associations. By increasing diversity locally, managers may unwittingly contribute to the local extirpation of birds and other species dependent on the mature forest that originally dominated the region. If such actions were repeated in other reserves, combined with the general loss and fragmentation of forests in the region, diversity could decline over a broad area as forest-interior species were gradually lost.

And this is not just a problem for birds. Studies have shown that the diversity of many kinds of plants and animals increases locally as weedy generalists, including many nonnative species, move into an area after disturbance by humans. In the case of Sugarcreek, the species that benefited when managers maintained an artificial interspersion of habitats were species common throughout the agricultural and suburban landscape. Many of these species would have been rare or absent in the undisturbed forest landscape. Management increased diversity locally by maintaining populations of the early-successional and edge species that had invaded the area during or soon after agricultural development. Similarly, clearcuts often show a pulse of diversity that rivals the diversity of the mature or old-growth forests they replace. This inflated local (and usually temporary) diversity is beguiling, leading some land managers to conclude that the more they disturb the land, the better.

Such is the fallacy of misplaced scale. Thinking locally leads to small-minded decisions and, ultimately, to global homogenization. Land managers need to consider a bigger picture, a broader context. They need to ask, What can I do on this piece of land that will contribute most to regional and global biodiversity over the long term? In this article, I explore this question in relation to three key conservation issues: enhancing population viability, managing disturbances, and prioritizing the protection of species and natural communities. The answers are not always straightforward, but the simple process of seeking them leads us to question established orthodoxies and confront new realities.

POPULATION VIABILITY IN A LANDSCAPE CONTEXT

Recent evidence suggests that the broader context of sites is meaningful in ways few biologists could anticipate 20 or 30 years ago. Every local area is a piece of a bigger ecological puzzle. Its importance can be appreciated only in relation to a larger whole. A natural area of seemingly marginal significance can be enormously valuable if its position and ecological role in the landscape contribute to the persistence of populations on a regional scale.

Over the last several years, my colleagues and I have studied the distribution of suitable habitats and the viability of populations of several species of carnivores—gray wolf, grizzly bear, black bear, wolverine, fisher, cougar, lynx, and others—in the Rocky Mountains of Canada and the U.S. Because many carnivores have demanding area requirements and occur in low densities, their viability must be considered over enormous areas. Our results, based on simulation models but validated by the fieldwork of other researchers across the region, suggest that the most vulnerable sites for carnivores are not necessarily those with the highest levels of site-specific threat (such as logging or residential development) but rather those whose continued degradation would affect nearby, high-quality areas that sustain regional populations.

The concept of population sources and sinks is germane here. A source population (or area) is one where the annual birth rate exceeds the death rate. Young animals born in sources disperse out to other areas, some of which are sinks. In a sink, deaths exceed births in a typical year, so a population can be sustained only by animals moving in from sources. Because individuals of a given species can be reliably found year to year in both sources and sinks, conducting research to differentiate the two is crucial to conservation planning and management. If a land-use plan only protected sinks while opening source areas to development, the population would quickly plummet to extinction.

Paradoxically, in some cases improving conditions in strong sinks might be nearly as crucial to regional population viability as protecting strong sources. For example, a heavily roaded area adjacent to Yellowstone National Park, where many grizzly bears are killed, may be a drain on the population within the park. The viability of the regional bear population could be enhanced by closing roads and limiting development near park boundaries, thus reducing the threat of mortality for animals dispersing from the park. Moreover, an area of low habitat quality for bears may be a key linkage between two high-quality areas whose connection is crucial because the population requires a combined acreage larger than either of the two high-quality areas alone. In this way, habitat connectivity can create a whole larger than the sum of its parts.

The point is to take a broader and more dynamic view of landscapes and populations. Most researchers and managers to date have relied on static habitat suitability models, which provide a snapshot in time of habitat conditions relative to a species' requirements. These models can be used to predict which parts of a landscape are most favorable for a species but tell us little about how easily a particular site might be colonized or how long a population might persist there. We are now realizing the utility of dynamic population models, which predict habitat occupancy and rate of population growth of a species over time. The dynamic models begin with data on habitat quality from the static models and then build on this information by simulating the birth, dispersal, reproduction, and death of individuals throughout a study region based on what is known about the life history of the species. Thus, the dynamic model provides a population viability analysis (PVA), but unlike many PVAs, it is spatially explicit.

My student, Carlos Carroll, applied these two kinds of models to carnivores in our studies in the Rockies. Using the dynamic model, he found that sites that appeared highly suitable on the basis of static models might not be occupied regularly by carnivores because they are too small or isolated. This was especially the case for the lynx, a species whose populations in peripheral areas depend on dispersal from areas of more continuous distribution. Whereas the boreal zone may be occupied consistently because of its high habitat quality and continuity—and the most southerly areas in the species' range are vacant for the opposite reason—areas in between may be occupied or unoccupied, depending on a complex combination of their habitat quality and proximity to source populations.

For the lynx and other species whose habitat is patchy as a result of either natural or human factors, much potentially suitable habitat is likely to be vacant most of the time. If we want to maintain these species in the more fragmented portions of their distributions, we must assure that human-created barriers do not add to the natural barriers that limit connectivity among areas of suitable habitat.

Birds, the best studied of all taxa, provide many other examples of the importance of considering landscape context and source-sink dynamics in conservation. In intensive agricultural regions such as southern Illinois and southern Ontario, most of the remaining patches of forest, and indeed the entire landscape, appear to be sinks for certain species of birds, such as the ovenbird and wood thrush. Predation by edge-loving predators, such as opossums, house cats, raccoons, and crows, and nest parasitism by cowbirds are usually the primary causes of reproductive failure. Why, then, are ovenbirds and wood thrushes still found in small forest patches in these regions year after year? Although proof has been hard to come by, it seems likely that the forest-interior birds that remain in these landscapes are dispersing in from other regions that are still dominated by forest. These regions, such as the Ozark Mountains and central Ontario, are population sources. The excess young birds move out, looking for new areas to raise a family, and wind up in sinks such as southern Illinois and southern Ontario. The only plausible way to change such regions from sinks back to sources is through massive reforestation.

DISTURBANCE MOSAICS NEED ROOM TO MOVE

One of my favorite ecosystems in North America is the longleaf pine forest that once dominated the southeastern coastal plain of the United States but has been reduced by 97 percent or more since European settlement. This forest (which is really a savanna—a grassland with scattered trees) was sustained by lightning-ignited fires that returned every two to five years or so. Today, managers of an isolated longleaf pine reserve might conclude that since lightning has not set a fire in their reserve for 15 years, then that must be the natural fire frequency, and there is no need to take management action.

Yet, if they were to step back and take in the bigger picture, they might draw a very different conclusion. The probability of a fire igniting in any single, small reserve is far smaller than the probability of an ignition somewhere across a vast landscape. In fact, in the unfragmented landscape that existed before Europeans arrived, a single ignition might have resulted in a fire that spread over many thousands of acres. Now, barriers such as roads, clearcuts, and urban areas prevent that natural spread—and hence the frequency—of fires.

All of the remaining stands of longleaf pine are fragments of varying size. Beginning in some of the more forward-thinking agencies, such as Florida State Parks, managers now regularly use prescribed burns to simulate the natural fire regime. Species that depend on the open-canopied stands of longleaf pine, such as the endangered red-cockaded woodpecker, have benefited from this management. Yet, it took managers a long time to recognize the broader spatial and temporal context of the remaining stands of longleaf pine and other fire-dependent communities. The general public does not yet see the bigger picture. Prescribed burning is now threatened in Florida and other regions because of public concerns about air quality and safety issues.

The need to consider context begins with the realization that, at one scale or another, all disturbances are patchy. In many kinds of ecosystems, natural disturbance and recovery processes create a shifting mosaic of successional stages across the landscape. In forests, newly disturbed sites, old-growth patches, and many stages in between coexist across a broad region. Over a span of time, the proportion of habitats in various stages may not vary much, although the actual locations of these stages will change relatively frequently as new disturbances arise and previously disturbed areas recover. In this disturbance mosaic, habitats for a range of native species that depend on different stages are maintained over time.

Disturbance mosaics, however, cannot operate in small areas. In a tiny reserve, a single intense fire or windstorm may eliminate all the old growth at once, leaving no refuge from which old-growth species can recolonize the disturbed area over time. Similarly, any earlier successional stage could disappear simply from lack of disturbance, and all the sun-loving species could be shaded out. Managers must not only consider the natural distribution of stages but also try to maintain some semblance of that mosaic through time and across the landscape as a whole, rather than within individual reserves or sites. In addition, given the broader historical and regional context of the landscape, managers should emphasize the development of those stages—typically old growth—that have declined the most since European settlement.

A recent study of fire history over the last 3,000 years in the Oregon Coast Range, combined with a simulation of changes in forest age classes over this period, determined that variability in the distribution of age classes over time was much greater in small areas—for example, the scale of an individual national forest or, smaller yet, a reserve within a national forest—than across the region as a whole. In a small reserve, old-growth or late-successional forest could easily disappear by chance, eliminated by a single fire. Across the entire region, however, the simulated proportion of old-growth forest varied mostly between 25 percent and 75 percent and averaged 46 percent, assuring dependable late-successional habitat conditions for such species as the northern spotted owl. When the proportion of old growth declines to a small level (currently about 10 percent) and persists at that level for a long time, the survival of associated species is jeopardized and their recovery is precluded. How much old growth is needed to restore viable populations of associated species is not known with certainty and would depend on the spatial arrangement of stands as well as their total area. Nevertheless, a prudent management goal would be to grow more old growth, up to at least 25 percent of the region, to bring it within the historic range of variability.

ARE ALL SPECIES CREATED EQUAL?

I am continually surprised by conservation assessments and management plans that, explicitly or implicitly, give all species equal weight. As living creatures, the European starlings and common grackles that I observed at Sugarcreek are every bit as respectable as the cerulean warblers and wood thrushes. Similarly, house flies and common dandelions are as admirable in the evolutionary scheme of things as the San Diego fairy shrimp or Ka'u silversword. But the former are not of conservation concern, whereas the latter are.

Perhaps all species are created equal, but treating all species as equal in any given place or time would be a disaster. As Jared Diamond noted nearly three decades ago, "Species must be weighted, not just counted... [as] the question is not which refuge system contains more total species, but which contains more species that would be doomed to extinction in the absence of refuges."

Although Diamond's advice is slowly sinking in, it has been an uphill battle. Consider diversity indices. In the 1960s and 1970s, these indices were common in the ecological literature and are sometimes still used today in environmental impact assessments. The simplest measure of diversity is species richness, the number of species in a defined area or sample. More complex indices usually combine richness with a measure of the evenness of abundances, with evenness peaking when all species in a sample have identical numbers of individuals. All these measures hide information on the identity of species, yet they have been used to place higher value on communities with higher richness or diversity, regardless of species composition. Thus, a species-rich community dominated by nonnative species would be favored over a community dominated by natives, but with fewer species. Although few conservationists make this error today, some people who propose engineering projects, timber sales, and other developments justify their projects on the basis that they will increase biodiversity, meaning the number of species locally, regardless of identity and conservation status. They create the same fallacy of misplaced scale that I saw at Sugarcreek, and some decision-makers still fall for it.

A useful tool for weighting species according to their conservation value is the "G/S" ranking system developed by The Nature Conservancy and now managed by a spin-off organization, NatureServe (see www.natureserve.org/explorer). This system tracks species according to their level of imperilment, based on rarity and other measures of vulnerability to extinction. The highest ranked species (G1) are considered critically imperiled globally. Most of these species are narrow endemics, found only within small areas and in small populations. One shopping mall could wipe them off the face of the earth.

At the opposite end of the spectrum, G5 species are demonstrably widespread, abundant, and secure. Species are also ranked on national

A Brief History of Thinking Bigger

In the 1960s, ecologists Robert MacArthur and Edward Wilson introduced their theory of island biogeography, which suggested that the number of species on an island or island-like habitat represents a balance between colonization by new species and extinctions of existing species on the island. These processes, in turn, are determined by how close the island is to the mainland (or to a large patch of habitat) and by the size of the island, respectively. Big islands near a source of colonists are predicted to be the most diverse. Although empirical support for MacArthur and Wilson's theory is mixed, and many other factors come into play in determining diversity, the theory was historically pivotal in that it got many scientists and conservationists thinking seriously about the effects of habitat area and isolation on population persistence and species diversity.

In the early 1980s, the field of landscape ecology was introduced to North America. Richard Forman and others discussed the factors that determine species distributions and abundances across areas many miles wide. They pointed out that the pattern of habitat patches and corridors—as well as the nature of the surrounding landscape matrix—could have a substantial effect on species composition and ecological processes. That now seems so obvious. Yet thinking about habitat at such broad scales differed strikingly from the scale at which wildlife habitat was usually evaluated in those days and often still is today.

In 1986, Larry Harris and I published a paper in which we urged conservationists to expand their thinking from the content of natural areas to the context of the landscape in which they are embedded. We noted that areas usually are selected for protection based on their contents—the species and communities they contain. Often the interest is in the scenic features or recreational potential of an area or a lack of conflicts with resource production. Sometimes it is just plain opportunism—the land is available for a reasonable price. Even when more scientific criteria are applied to selecting reserves, for example, representing a particular plant community or protecting a population of a rare species, the focus is on the piece of land or water itself, not the landscape that surrounds it. Hence, the native biodiversity of many reserves degrades over time as area-demanding species drop out and external influences increase.

Conservation planning today, in contrast to just a decade or two ago, is usually pursued on a regional scale. The Wildlands Project, formed in 1991, pioneered the development of regional reserve networks, drawing on preliminary designs I and a few others developed during the 1980s. Ultimately, the Wildlands Project plans to extend such designs across North America and beyond. The U.S. Gap Analysis Project, initiated in the early 1990s by J. Michael Scott and Blair Csuti and now housed in the U.S. Geological Survey, set out to evaluate how well biodiversity is represented in existing reserves across the country. The inevitable answer was—not too well. By the mid-1990s, The Nature Conservancy, World Wildlife Fund, Conservation International, the Wildlife Conservation Society, and many other groups had embraced ecoregional planning or similar large-scale efforts. Meanwhile, Australian biologists, notably Robert Pressey and colleagues, developed sophisticated computer algorithms that made the selection of new reserves to complement existing ones much more efficient. Given explicit goals and reasonably good databases on biodiversity surrogates, these algorithms can answer the perennial question—how much is enough?—much more reliably than in the past. Finally, we are beginning to see the big picture.

and subnational (e.g., state) scales, and the system is similarly applied to ecological communities. A species or community that is found in only one site in a state, but is common elsewhere, might be ranked G5S1. A species or community with this rank usually would not warrant as much concern as one ranked G1S1 or G2S1, but it would still be worth protecting, especially if it were a disjunct population that is genetically distinct from the rest of its species.

Today, conservationists and managers are more concerned with entire ecosystems than with single species, and rightly so. But here again, simplistic notions of diversity are foolhardy. Maximizing the diversity of ecosystems in an area through management makes no sense. The historical as

Six Lessons on Thinking in Context

1. The integrity of any piece of land or water is ultimately dependent on the integrity of the landscape that surrounds it.
2. Many species require areas well outside protected area boundaries. Whenever possible, work with regional planners to assure that key pieces of habitat are functionally connected, through habitat corridors or low-intensity intervening land use.
3. Ecological history is highly informative. Which habitats and species have prospered and which have declined as the extent and intensity of human activity increased?
4. Bigger scales are ultimately more meaningful for conservation of global biodiversity. Recognize that management may increase the number of species on one spatial scale while decreasing it on others.
5. Do not treat all species as equal. Focus attention on those most vulnerable to extinction as a result of human actions, those that have declined most, and those that are ecologically significant (including exotics and other opportunistic species that threaten native biodiversity).
6. Limit human activities around and within reserves and other natural areas. Surround sensitive areas of any size with transitional or buffer zones where the intensity of human uses increases with greater distance from the sensitive area.

well as the spatial context of ecosystems must be considered. Ecosystems in the United States that have suffered dramatic declines since European settlement—for example, longleaf pine and coastal redwood forests, tallgrass prairie, and seagrass meadows—deserve greater attention than ecosystems that are still relatively common, and arguably more than those ecosystems that have always been rare. Treating different species and ecosystems as nonequals might smack against our egalitarian ideals, but it is the only efficient way to confront the extinction crisis.

KEEPING TRACK

RICK BASS

It's five below zero when I leave Concord, New Hampshire, snow scrunching underfoot and stars flashing, and drive north and west into Vermont, driving through the night toward a visit with Sue Morse and the tree farm that she caretakes. She is a professional tracker and a forester, and will be leading a class the next day on a walk through her wild woods, just south of the region known as the Northeast Kingdom. Sue lives in country that wolves will probably one day reinhabit, not simply because human beings desire it, but because it was once partly their country before. The shape of the land is still the same, has not changed much at all since the last glaciers; and once again, like a tide, wolves will slide back down over the hills and mountains and through the forests, laying down tracks upon their own ghosts.

Driving through the hills, up knolls and swales, through the snowy darkness, along the winding back roads past sleeping hamlets, it's hard to imagine even a single wolf navigating its way secretly through this countryside. But perhaps it could. Perhaps if it kept traveling, it could move through these sleeping woods with care.

Sue Morse radiates something; I'm not sure what. You could call it health, strength, vigor, even wildness. Whatever it is, it's vital. Her short strawberry-blond hair is burnished as if from much sun, and in the thirty-below wind chill this crystal-bright March morning, her face is ruddy, which gives her broad smile extra cheer. It's clear that this morning's joy comes from the prospect of getting to go out-of-doors today, weather be damned, to simply be in the woods. It is always a great and specific joy to be in the presence of one who finds such good cheer in the midst of—because of—harsh or inclement weather.

Sue takes people on woods walks, and teaches them tracking, throughout the year. She's helping coordinate a training program in Vermont called Keeping Track, wherein volunteers conduct wildlife inventories in the woods around their homes. Across the years, these inventories will provide the state with a bank of reliable biological data and trends that could not otherwise be obtained.

Sue passes out walking sticks to the couple dozen participants, and we start up the mountain, shuffling through knee-deep snow. The forest is covered with snow from the day before, glaringly brilliant in the morning sun, and the sky possesses that depth of blue that you see only against new snow, and only at temperatures well below zero.

Because she brings so many people on these walks over the course of a year, Sue asks that we take extra care to stay in single file and, where possible, to stay in each other's footprints, to keep from bruising or trampling dormant plants and seedlings beneath the snow.

This isn't to say that humans haven't been active in this forest. Sue and her neighbors have logged portions of these woods selectively, with care and pride, always toward improving, as they can, the health and vigor of the aging forest. She has a hunter's eye for which trees to leave, which to take, and which to let grow a bit older, just as a hunter might pass on a certain deer, even a nice deer, as long as the deer was still in its prime and growing vigorously. (Sue is an avid muzzleloader.)

She also recognizes the full value of a dead or dying tree, an aging giant, to be left in the woods forever. She knows her forests intimately—the comings and goings of the seasons, the curves of the land and the qualities of the soil in each little fold and pocket.

It gives me solace to be in her company, to be walking along behind the aura of her strength and confidence, her passionate moderation. So often, when I am seated at the computer, pounding away at the keyboard and feeling under the activist's siege, I feel that I am becoming evermore a lightweight, always bending over backward to take into account the opposition's position, considering their opinions and perspectives. I'll find those concerns factoring into my every thought, and because it feels like battle, and because I do not want battle, I'll often feel as if I'm being too radical—even when I'm writing what sounds to me like the sanest, most reasonable things. I feel like an extremist.

But then, whenever I get out into the woods and get up on some windy ridge, and look out not at some computer screen but at the rise and fall of the landscape I know so intimately—such a small piece of the west, but so important—always, always, I'll feel a thing like anguish: not that I have been too radical, but that I have been too moderate, in the face of what is at stake, and the dangers surrounding and encroaching on these last wildlands.

Listening to Sue talk about her hunting experiences and her belief that we can still have "wilderness *and* logging, rather than wilderness *or* logging," inspires me. Her confidence, as well as the authority of her experience, inspires me to believe that the middle path *can* be had, just barely, if we hurry, and that we can still save the last corners of untouched, unprotected wilderness and repair the injured lands in between.

It's a gamble: cutting trees to save them, cutting trees to save forests, cutting trees to try to save threatened and endangered species, rather than simply committing to a 500-year plan of hands-off and stepping back to let things sort themselves out with nature's customarily intricate and unfathomable grace. But it's a gamble I believe in. It's a pleasure to find the company of another who feels the same way.

It's astounding, really, to meet someone who is so similar to one's self in so many ways, and yet who seems upright in the world—so unfloundering.

Is the east only a less frantic reflection of the west, or the west a more panicked reflection of the east, as more and more of the west's wild treasure is spoiled, ruined, as the east's once was? (And what irony, too, that as Congress continues to urge onward the liquidation of the public lands of the arid west, the lusher, more resilient east is recovering; there now exists in northern New England alone more than 26 million acres of forested wildland.)

East or west, for my part, I can only laugh. The face in the mirror over here isn't the same, but nonetheless, other mirrors seem to be reflecting the same hungers and desires.

Sue Morse loves to build stone walls! She can be fierce and ornery with "the opposition," though infinitely patient with the students she's teaching. She loves books. She loves feathers, stones, driftwood, hides, bones: the physical beauties that anchor this world of spirits, the structures and frames for the sparks within.

How different, how individual, really, can any of us be from one another—especially when we love the same things? The forests and the fields, beaches and deserts of the world are far more individual than our various hearts can ever be. When we're lucky, we are but echoes and shadows of those landscapes. When we have destroyed them, our own lights will cease to exist; we will vanish, cut off from the thing that had for all the centuries and millennia before mirrored and shaped us, and shaped our hearts.

We haven't traveled far before we hit the first tracks; another living thing has passed before us. A young coyote has been scavenging a deer carcass, and Sue shows us its tracks. Loose deer hair is scattered here and there, and newly frozen scats puncture the coyote's wandering sentence of tracks, scat wrapped also in deer hair and studded with pearls of jaw-cracked bones.

In an open area, Sue shows us other sentences that pass over and across the coyote's sentences, the feather-swept strokes where a raven had swooped in for a bit of bone or flesh, or perhaps to visit the coyote.

I don't want to dwell overmuch on the notion of spirits—such thoughts and discussions almost always wither when inked onto paper, and are instead best explored, I think, in private caldrons and eddies of each person's heart—but I can see how for Sue Morse, having the knowledge, experience, and ability she does, it might be a little like being able to look into the spirit worlds in motion all around us.

As she moves though the woods, noticing the residue of tiny and not so tiny passages of others, it must be for her like a kind of second vision. She is able to see the hidden past as clearly as you or I might were we to take a few steps into the future and then look back to view the present from some slight distance ahead.

Not much farther beyond the coyote tracks, we find fisher tracks. In the soft, deep snow, the prints are not very distinguishable. But the distinct, loping, sinuous carriage of the animal—only the wolverine is larger among the mustelids—identifies it. Sue makes a few hops of her own, imitating for us that weasel-like swimming motion with which the mustelids attack the world, as if part fish, part snake, part mammal.

Deeper still, we find bobcat tracks. The sunlight in the old interior forest of leafless beech and maple is dense, seeming somehow from another time; columns of buttery gold slant against the winter-pale bark of the trees, mixing with the deep shadows to create a yellow and black mosaic that is nearly identical to the coloration of bobcats; and again, it is like a kind of vision, seeing those tracks pass through that light and knowing that we are only hours or even minutes behind the real animal.

Sue even knows this bobcat by name; she knows its patterns, it reproductive history, its size and temperament. She has been hiking these woods almost daily for the last twenty-three years.

Loosely formed as they are, the tracks are hard to distinguish from those of a coyote. Sue has us follow the tracks backward—she doesn't like to push an animal ahead of her, is impeccably polite and courteous that way—and finally we come to a spot of diagnosis, a spray of fresh urine, which she instructs us to kneel down and smell. *Cat.*

In a recent *Audubon* article by David Dobbs, Morse explained the strategy behind her Keeping Track classes: There is scientific as well as activist value in having citizens gathering data from their home terrain throughout the state, recording the patterns and occurrences, the comings and goings of things, with a degree of coverage that no state or federal budget could ever afford.

"Only by doing this sort of homework—literally keeping track of wildlife use—can we know a given habitat's true value," she said. "And only by knowing its value can we work effectively to conserve it."

The place where we're walking is, she says, a case in point. "Twenty years ago, if we had discussed as a town what to do with these lands, [all] I would have said was, 'Whoa! We can't build a Wal-Mart here! These woods are too pretty! I've seen moose here. This is special!'"

Now, however, with the weight of her documentation behind her, she can say, "The area functions as core habitat for bear, bobcat, fisher, mink, moose, and otter, and it's essential to the health of those species and everything they draw on. It's a big difference."

It's brutally cold back in the shade of the forest. Everyone is flapping their arms and gloved hands like stunned chickens, laboring unsuccessfully to stay just this side of suffering—it is only the beauty of the old forest that allows us to push back the pain in our toes and fingertips even remotely. Amazingly, Sue Morse is striding along bare-handed. She's breaking trail for us, so I'm sure her blood is running a good bit harder and hotter, and you can tell just looking at her that she's thrilled, always thrilled, to be in the woods. But the wind chill is still way below zero, and I can't help but marvel at all the different tones and variations of wildness in the world, including among our own kind.

We come out of the old bottom, ascend a ridge and begin to find more bear sign. The great beech trees have been climbed and clawed, the smooth bark latticed with the ancient hieroglyphics of bears past and present. Sue pointes out the ladder-stacked clusters of branches in some of the higher trees: large nests that the bears have pulled together by climbing out as far as they can on limbs and then reaching for the smaller branches, breaking them off and wedging them in sturdy crotches of the trees. There the bears strip off the nuts and fruits, eating them the way you or I might pluck grapes from vines.

We hike all day, crossing the occasional tracks of bobcat, otter, fisher, deer, ascending cliffs and descending along rushing creeks, until it seems certain that we have left Vermont and are in some other, wilder place.

Nearing the end of our loop, returning to Sue's cabin, we pass along the boundary line of the property she tends, which is punctuated by immense, ancient maples—"witness trees" that were not cut but blazed to serve as survey markers—trees that were substantial even back in the days of the land grants, 200 years ago.

We pass the rusting hulk of a crumpled wood stove, moss-anchored, being swallowed by the forest. At the beginning of the century, Sue tells us, almost all of the forest we're walking through was cut-over fields and denuded sheep pastures. There were accounts of wolves being so hungry back then that they would come out into the meadows and gnaw on the maize from farmers' fields. Then, like the forest itself, the wolves went away for a while.

Now the woods are coming back, and the wolves will be coming back, and barely a single human lifetime has passed, two humans' lifetimes.

Our little cycles are nothing, compared to the world's. Our greatest dramas are like those of the mayfly. We spend a few weeks scrounging gravel and building a little edifice. Then some force summons us—we ascend to warmth and light, spin with violent beauty and a strange mix of joy, confusion, instinct, and desire—then fall from the sky, back to the river, or to the gravel bar at river's edge.

Farther downstream, all around us, other mayflies rise, like the lifting and falling of the world's threadlike pistons. As if, daily, all disappears, is buried, is lifted up, returns.

That night, we dine in Sue's cabin, which is filled with more books than I've ever seen in a cabin. The stuff of her life surrounds her—spotting scopes, telescopes, binoculars, hides, pelts, antlers, rocks, feathers, and bones—all beautiful, testimony to the great grace and design of the world. But what strikes me most is the books: shelves of books, books on tables, and books resting in chairs, half-read. What magnificent paradoxes we are, for a woman as fully engaged in the out-of-doors as Sue Morse is, to also have her life, her spirit, so immersed in ideas, the invisible and abstract concepts of words and sentences, each of which always drifts a slightly different path for each reader, rather than possessing the clarity of a track in snow or a claw mark upon a tree.

We dine on hot lamb chops and cold beer, sitting cross-legged around the wood stove, surrounded by those great books of poetry and natural history, novels and field notebooks, journals and scientific books; and we discuss, as it seems environmentalists are always doing, the future.

Sue raises organic, free-range sheep on her farm. We're eating one of her lambs tonight. She doesn't trap coyotes, but instead uses guard dogs to protect the sheep. She wants the wolves to come back. It seems at first a wild and wonderful paradox, her wanting both sheep and wolves, but then, after a day in the richness of her country, it comes to seem natural. There's still—with resolve—barely time and space for both.

It's her job to take care of the sheep; it's her job to take care of the wolves. *Stewardship.*

The scholars, professors, and activists Chris Klyza and John Elder have joined us. Chris, an intense and intelligent young man, talks literature for a while, until, like magma seeping from vents in the earth, numbers and statistics instead begin to issue from him. The fact that 20 percent of our population is already experiencing a shortage in fresh water supplies, and that in the next twenty-five years, more than 30 percent will be lacking or stressed.

The fact that the world's human population is expected to double in the next fifty years.

The fact that 60 million people live within a day's drive of where we're sitting.

Are we—moderate environmentalists—blind to the reality of the situation? Are we blinding ourselves, perhaps as a defense against the awful weight of the truth, while we move through the world neck-deep in our own ferocious consumption and continue to proclaim that hope for the future still lies in words like *sustainability, stewardship,* and *moderation*—that indeed, we can (and must) have more wilderness and, at the same time, more (but smarter) logging?

Can we still find a middle way to carry forward these two desires, these two needs, rather than abandoning one or the other?

Probably, for one or two more years. This is probably the last cusp of time in which an environmentalist can be afforded the indulgence of believing such things. Surely it is our last last chance.

John Elder agrees. "The game is now," he says. He's talking about Vermont, and the rest of New England. "The pressures to develop are going to become even more intense in just the next few years," he says. The booming economy, the astounding affluence, the eagerness of the large landholding companies to divide their land into the smallest possible fragments. . . .

Where are the wolves in this story? They're still a couple hundred miles north, still out in the future. An unconnected triangle still separates wolves and this landscape, and the desire in our hearts for the wolves to move back onto this landscape.

But it seems that the pieces are beginning to move closer together, almost like a thing that is getting up and beginning to walk.

And later that evening, as we're talking about innovations in forestry practices, value-added products, sustainability, the night's frigid silence is broken by a coyote's howl.

It's so close and so loud that we can hear it even through the walls of the cabin, through the old logs from the forest, through the closed windows.

"It sounds like a *wolf*," I say, and I look at Sue with some suspicion, wondering if she's not keeping some sort of secret.

"The coyotes are really big around here," she says. "They can weigh up to seventy pounds" (the size of a young or small wolf).

The four of us wander out into the depth of the night. It's like walking into a dark ocean. John and Chris are marveling at the stars. We truly could be at the edge of a wilderness. Places like this, pockets and patches, still exist in the east.

The coyote's howls are wild and fierce and strong: *so close.*

One of us murmurs something and the coyote hears it and falls silent. We wait for it to resume, but it's too smart, too wily. For us, it's too cold outside; we linger as long as we can, a few moments longer, stamping our feet and waiting and listening.

Sparks and the smoke of burning trees rise from Sue's chimney, up into the leafless canopy of the trees; higher still, then, the sparks seem to be drifting through the stars themselves.

We stand a few seconds longer in the icy night, just outside the glowing yellow cabin, caught perfectly between domesticity and wildness, as if that's where we're most meant to belong. If so, when there is no more wildness, or no more wilderness, what shall we do? What will become of us?

It's so cold that we can see the long crystals of ice slashing and glimmering in the starlight as they mix in a slight stirring breeze. They look like teeth, long sharp teeth, falling upon the land, gnawing and tearing at the land, but causing no damage.

THE ROLE OF TOP CARNIVORES

JOHN TERBORGH, JAMES ESTES, PAUL PAQUET, KATHERINE RALLS, DIANE BOYD-HEGER, BRIAN MILLER, AND REED NOSS

The role of predation has become a matter of intense interest to conservationists because mounting evidence, as we shall see, points to its pivotal role in helping to preserve the biodiversity of terrestrial communities. On every continent, top predators are now restricted to tiny fractions of their former ranges, so that the integrity of biological communities over large portions of the earth's terrestrial realm is threatened by grossly distorted predation regimes. Even where they are present, top predators' population densities tend to be so low, and their behavior so secretive, that sightings are infrequent. Most biologists prefer to study species that are common, small, and easily manipulated. Many academics dismiss field studies of large carnivores as "unscientific" because sample sizes are typically small and controlled experimentation difficult. Carnivore biology has thus been left to a small coterie of hardy devotees whose work, if not ignored, lies well outside the mainstream. The role that top predators play in terrestrial ecosystems, therefore, remains ill defined and contentious. Here, after reviewing a large body of literature, we conclude that the loss of top carnivores from their native ranges may cause a cascade of ecological effects that speeds extinction.

Whether contentious or not, it is crucial to define the role of top predators—the stakes are enormous. If, as we conclude here, top predators are often essential to the integrity of ecological communities, it will be imperative to retain or restore them to as many parts of North America as practical. Failure to do so will result in distorted ecological interactions that, in the long run, will jeopardize biodiversity across the continent.

THEORY

What is at issue in the current debate over "top-down" versus "bottom-up" processes? "Top-down" means that species occupying the highest trophic level (top carnivores) exert a controlling influence on species at the next lower level (their prey) and so forth down the trophic ladder (or food chain). The definition can be made operational in a thought experiment. Under top-down regulation, the removal of a top predator (or better, the entire suite of top predators) results in an appreciable population increase in the prey. It is thereby demonstrated that productivity (the food supply available to the prey) was not the factor limiting prey numbers. Conversely, if removal of the top predators does not lead to increases in the numbers of prey, we must conclude that the prey were limited by something else—most likely the food supply.

We can ask parallel questions about the bottom rung on the trophic ladder. Suppose we could increase the longterm productivity of an ecosystem experimentally—let us say by adding water to a desert or nutrients to a barrens. If the increase in plant growth resulting from the artificial input then led to an increase in the biomass of consumers (herbivores such as rabbits and deer), we could conclude that the consumers were under bottom-up control. If we found no increase in consumer biomass, this would imply that something other than plant food was limiting—plant defenses, or predators, to mention two possibilities. Even by the admittedly simple operational criteria just presented, it should be evident that top-down versus bottom-up is not merely an either/or proposition.

Both top-down and bottom-up regulation can operate concurrently in the same system. In the presence of predators, herbivores are secretive and endeavor to spend as little time feeding (when they are exposed to predators) as possible. Most of their time is spent in secure places—in burrows or dense thickets, for example. If predators are removed, the quest for security ceases to be the leading regulator of prey behavior; now consumers are free to feed when and where they want, becoming energy-maximizers, also maximizing the ability to reproduce. The switch in prey behavior in response to differing levels of perceived predator threat introduces complexity into the system and allows both top-down and bottom-up regulation to operate simultaneously or to varying degrees. A plant's defenses add another layer of complexity. Damage to foliage can stimulate plants to increase levels of herbivore-deterring chemicals in their tissues—thereby reducing the food supply available to herbivores (a bottom-up effect). It is the extraordinary complexity of food interactions that makes the issue of top-down versus bottom-up a matter of so much contention among ecologists.

Top-down effects have been shown to act on communities in two fundamentally different ways. One is in the absence of predation a prey species is capable of competitively excluding other species that depend on the same limiting resource (such as food). Thus, over an intermediate range of predation intensities, species diversity of prey is enhanced when predators are intermediately abundant (midway between absent or overabundant). This process, known as the intermediate disturbance model of species diversity or the Paine effect, has been demonstrated in a variety of systems.

The second way in which predators influence their communities is through a cascade of interactions extending through successively lower trophic levels to autotrophs at the base of the food web. In trophic (food chain) cascades, for example, plants do better when a reduced number of herbivores are feeding on them or worse when herbivore numbers increase. The top-down model predicts that each trophic level is potentially limited by the next level up. For intact three-level systems, therefore, predators limit herbivores, thus releasing producers (plants) from limitation by herbivore feeding.

EMPIRICAL FOUNDATIONS: THE INTERMEDIATE DISTURBANCE MODEL OF SPECIES DIVERSITY

If terrestrial carnivores were not so inherently difficult to study, we might have understood their roles long ago. Simpler conditions characteristic of certain aquatic systems have facilitated investigation, however, and the keystone role of predators is now established beyond dispute. Paine was the first to provide incontrovertible evidence. By removing the predatory starfish *Pisaster ochraceous* from sections of the intertidal zone of the rocky Washington coastline, he showed that the diversity of the attached invertebrates subsequently declined as a superior competitor, the mussel *Mytilus californicus,* gradually occupied all available space, thereby excluding other species from the community. It is important to note that *Mytilus* is the preferred prey of *Pisaster,* so that the action of the predator is selective removal of the dominant competitor—an act that exposes attachment sites that can be exploited by other species. Further studies of sessile intertidal communities have amply supported Paine's result. The primary effect of a top predator in the intertidal system is thus seen in regulating the diversity of the prey community. This is the Paine effect.

The presence/absence of a predator influences the productivity and biomass of the intertidal prey community because space (attachment sites) is the limiting resource. Terrestrial and aquatic systems involving mobile organisms may show different dynamics, however, because consumers and predators are free to come and go and many of the component species have long lifetimes. And unlike Paine's rocky intertidal system, which can be studied on the scale of a few square meters, terrestrial and open-water aquatic systems must be studied on vastly larger spatial scales because the important predators and consumers may have low population densities and range over large areas. These research obstacles have been major impediments to scientific progress. Now, with results emerging from some long-term studies and the first large-scale predator-exclusion experiments, the time is ripe for a synthesis.

ANECDOTAL EVIDENCE

In the hope of arriving at some general conclusions, we now review evidence relevant to understanding the role of top carnivores. Our emphasis is on terrestrial ecosystems and large vertebrates, especially mammals. The evidence is broadly categorized as anecdotal or experimental.

HERBIVORE RELEASE ONTO PREDATOR-FREE ISLANDS

Sailors of yore introduced horses, cattle, sheep, goats, pigs, and rabbits to predator-free islands throughout the Seven Seas to ensure themselves of a supply of meat on subsequent voyages. Few of these introductions were carefully monitored, so they can hardly be considered scientific studies. Nevertheless, in numerous instances the introduced herbivores increased without check until they devastated the native vegetation of the island—at which point populations of the herbivores themselves often crashed.

Destruction of the vegetation on predator-free islands by herbivores could be a top-down effect, but another interpretation is possible. The vegetation of islands lacking native vertebrate herbivores might have plants without herbivore defenses. Without additional information, we cannot distinguish the two interpretations, and both may be correct.

PREDATOR ELIMINATION

Humans have eliminated top predators over much of the globe, drastically reducing the geographical ranges of many species, including wolves, bears, tigers, lions, and many less intimidating beasts. Nevertheless, herbivores generally have not overrun predator-free portions of the planet, as we would expect if herbivore populations were indeed under top-down control. The reason in this case appears obvious. Large vertebrate herbivores are also the prey of human beings, and in many places they have been reduced to low densities or extirpated by human overhunting. In many regions, introduced livestock substitute for missing native ungulates. Untangling the effects of predator removal from those of hunting and introduced livestock is an almost impossible task in most situations.

One common situation that conforms to the proper test of top-down control is found in U.S. suburban areas where hunting is now prohibited. Mammals that would have been taken by wolves and cougars have become notoriously abundant to the point that some of them are now nuisances: by being road hazards (deer, moose); by browsing ornamental shrubbery (deer); by raiding trash cans (opossums, raccoons); by preying on birds (house cats) and their nests (cats, raccoons); by destroying vegetable gardens (deer, woodchucks, ground squirrels); and by flooding people's yards (beaver).

PREDATOR INTRODUCTION

Another kind of uncontrolled experiment is performed when predators are intentionally or unintentionally introduced (or reintroduced) into predator-free environments. The recovery of the sea otter from near extinction is a classic example. Sea urchins, abalones, and other benthic grazers had nearly eliminated the kelp forests that once dominated the inshore environment along the Pacific rim of North America in the absence of sea otters. Gradual recovery of the sea otter during the mid-20th century has led to sharp declines of benthic grazers, accompanied by dramatic recovery of kelp forests and associated fauna.

The introduction of exotic predators to predator-free islands provides additional evidence for the operation of top-down regulation. Mongooses introduced onto islands of the tropical Pacific and Antilles have contributed to the collapse of native faunas. Inadvertent introduction of the brown tree snake onto Guam led to a population explosion of the snake and consequent extinction of most of the island's native birds. Introduced domestic cats have had strong effects in Australia and on certain temperate islands, as have foxes in boreal to arctic regions.

On the North American mainland, the growing gray wolf population has been associated with a concurrent decline in elk and white-tailed deer densities. The recent reinhabitation of the northern Midwest by wolves has reduced the distance from aquatic habitats that beavers can forage—a behavioral modification that in turn reduces the impact of beaver on plant associations. Similarly, the reestablishment of wolves in other areas has been followed by declines in caribou, moose, elk, and deer.

LONG-TERM MONITORING OF PREDATOR/PREY INTERACTIONS

A compelling case for a terrestrial trophic cascade is that of the gray wolf/moose/balsam fir interaction on Isle Royale, Michigan. The number of wolves determines the intensity of wolf predation on moose populations on Isle Royale. Growth rings in young fir trees showed depressed plant growth rates when wolves were rare and moose abundant. Broad ramifications within the forest ecosystem are suggested from known linkages among moose, microbes, and soil nutrients.

The anecdotal evidence cited here is consistent with top-down regulation as a predictable feature of terrestrial and many aquatic communities. But without rigorous controls, anecdotal evidence, by its nature, is open to alternative interpretations, so scientists put greater stock in controlled comparisons and experiments.

EXPERIMENTAL EVIDENCE

Few well-controlled comparisons of prey populations at sites with and without top predators have been made—presumably because the conditions required are so rarely available. One carefully documented comparison is between two sites in the neotropics: one is Barro Colorado Island (BCI), Panama, a research preserve of the Smithsonian Institution; the other is Cocha Cashu Biological Station (CCBS) in the Manu National Park of Perú. Located respectively at 10° north and 12° south latitude, the two sites have a similar climate and fauna. The dominant habitat at both is primary tropical moist forest. BCI is a 1600-hectare island created by flooding during the construction of the Panama Canal. It has been isolated since the canal's creation. Due to its limited area, BCI lost top predators—jaguar, puma, and harpy eagle—more than fifty years ago. CCBS is located in the heart of a two-million-hectare biosphere reserve that retains an intact flora and fauna, including all top predators.

The terrestrial and arboreal mammals on BCI show higher densities than at CCBS. In several cases, the differences in abundance are striking—exceeding an order of magnitude, particularly for the agouti, paca, armadillo, and coatimundi (terrestrial) and the three-toed sloth and tamandua (arboreal). Differences in abundance are most pronounced in medium to large species that are prey of the top predators missing from BCI. The higher densities have been interpreted as evidence of a top-down effect resulting from missing top predators. Other interpretations are possible, however, including uncontrolled differences in productivity between the two sites.

The only certain way to exclude possible influences of uncontrolled variables is with strictly controlled experiments that include censusing before and after. There are now two experimental efforts underway that promise to overcome certain weaknesses. Neither set of experiments is perfect, but both represent major advances over previous efforts to isolate the effects of predators on terrestrial communities.

The creation in 1986 of one of the world's largest hydroelectric impoundments—Lago Guri—in the Caroni Valley of east-central Venezuela has resulted in the inundation of a hilly forested landscape with the consequent isolation of hundreds of erstwhile hilltops as islands. Islands ranging in size from less than one hectare to more than 1000 hectares are scattered throughout the vast expanse of water—a number of them as far as seven kilometers from the mainland. Small size and isolation by water assure that many of the more remote islands in Lago Guri are free of vertebrate predators except for certain small raptors and, perhaps, snakes.

Systematic surveys have shown roughly 75 to 90 percent of the species of terrestrial vertebrates that occupy the same forest type on the mainland were absent from islands between one and ten hectares in size within seven years after isolation. With few exceptions, species that persisted became hyperabundant compared to their densities on the mainland. The absence of many species and the hyperabundance of others has created animal communities unlike any that would ever occur naturally—communities that are grotesquely imbalanced from a functional standpoint. These communities lack vertebrate predators and are deficient in pollinators and seed dispersers; but they contain abnormally high densities of seed predators (small rodents) and generalist herbivores (howler monkeys, iguanas, and leaf-cutter ants).

Larger Lago Guri islands (between 100 and 1000 hectares) still retain nearly complete vertebrate faunas (all primates and ungulates known for the region, for example), lacking only resident populations of the top predators (jaguar, puma, harpy eagle). Mammal densities on the two large islands being monitored have not yet increased conspicuously, but one and perhaps both of these islands are visited regularly by jaguars that swim over from the mainland, so they are not strictly predator-free. As for the smaller, more isolated islands that assuredly are predator-free, the hyperabundance of persistent vertebrates is consistent with the top-down effect of release from predation.

Finally, we come to the most carefully constructed test of top-down regulation conducted to date—the monitoring of snowshoe hare populations in one-square-kilometer plots in southern Yukon, Canada. Two of the plots are surrounded by electric fencing that excludes large mammalian predators but is permeable to hares. Plots have been assigned to five treatments: control, food supplementation, fertilizer, predator exclusion, and predator exclusion with food supplementation. Averaged over the peak and decline phases typical of snowshoe hares, hare density was double that of controls under predator exclusion, triple with food supplementation, and eleven times greater under predator exclusion coupled with food supplementation. The results strongly implicate both bottom-up and top-down regulation. This interpretation is complicated, however, by the free passage of hares in and out of predator exclosures and by the exposure of hares within exclosures to predation by goshawks, great horned owls, weasels, minks, wolverines, and fishers. Nevertheless, the effort represents a bold attempt to conduct an experimental test of bottom-up and top-down regulation on an appropriate spatial scale with a natural predator/prey system.

Another series of large-scale experiments has been conducted to test the role of top-down regulation in freshwater aquatic systems. Entire lakes in Wisconsin have been seined free of piscivorous (carnivorous fish) or planktivorous (herbivorous) fishes and the respective hauls exchanged between lakes. Removal of piscivorous fish (large-mouthed bass, the top carnivore in this system) leads to order-of-magnitude increases in planktivorous fish, decreases in the size and number of zooplankton, and strong increases in the standing crop of phytoplankton in a textbook top-down trophic cascade.

Admittedly, many questions remain to be answered by future research. Nevertheless, with so much evidence pointing in the same direction, the conclusion that top predators play a major regulatory role seems inescapable.

COUNTERCURRENTS

Although the evidence that top predators commonly limit the densities of their prey is compelling, one would be wrong to conclude that predators limit the numbers of all consumers. There are a variety of situations in Nature that allow consumers to escape predation to varying degrees including both very large herbivores and herd-forming migratory ungulates. Nearly all the earth's once-abundant megaherbivores have been driven to extinction and only a few survive. In Africa there are rhinos and hippos, in addition to elephants, which, as adults, enjoy immunity to lions. In the north, adult moose repel gray wolves; in the neotropical forest, tapirs shrug off jaguars. Elsewhere, Madagascar had its elephant birds, New Zealand its moas, the Antilles their hutias and ground sloths, and the Seychelles, Galápagos, and Aldabra Island their tortoises. Lacking any population control from the top, megaherbivores must be regulated from below. But to the extent that megaherbivores regulate vegetation, they too exert a top-down force that is independent of predation. Ubiquitous and abundant to the point of dominating mammalian biomass over most of the globe for millions of years, megaherbivores have been so systematically persecuted that they have become almost irrelevant to today's ecosystems and conservation concerns, except in dwindling portions of Africa and Asia.

Size is not the only successful anti-predator strategy to have arisen through evolution. Some species are able to reduce (but not eliminate) predation through social mechanisms. The list of these mechanisms is long. It includes the formation of herds and flocks, sentinel behavior, and the giving of alarm calls. Consider the fabled wildebeest of Serengeti. These antelopes aggregate in huge mixed herds that can be within the territories of only one or two lion prides at a time. Lions only kill about 8 percent of the population per year. In a bad year, wildebeest die en masse from starvation and malnutrition. The conclusion follows that wildebeest—and, by analogy, other herd-forming migratory ungulates—are regulated from the bottom up. But again, how much of today's Earth is occupied by herd-forming migratory ungulates? Not much more than is occupied by megaherbivores. Both of these major agents of top-down forces in terrestrial ecosystems are becoming Pleistocene relics. Hence we should give special attention to top carnivore processes, because it seems likely that they are crucial to preserving what bits and pieces of wild Nature we have left.

Top predators play structuring roles in many ecosystems. Exceptions, however, may be found in extreme environments, such as deserts or barrens, where low plant productivity or chemical toxicity of foliage limits large herbivores to such a degree that predators are unable to exploit them. Other factors, such as a severe disturbance, can temporarily upset normal relationships. A stand-replacing fire, for example, may result in lowered herbivore densities and a switch from top-down to bottom-up regulation until the vegetation recovers. In the world at large, however, productivity-limited (pure bottom-up) systems appear to be rare. Moderate to strong top-down regulation appears to be the norm for terrestrial ecosystems.

INDIRECT EFFECTS AND TROPHIC CASCADES

Having made a case for top-down regulation as a nearly ubiquitous force in terrestrial ecosystems, we now ask about the role played by top predators in maintaining ecosystem integrity. What are the destabilizing forces that are unleashed in ecosystems from which top predators have been eliminated? What are these consequences and how severe might they be?

The intellectual groundwork for studying "indirect effects" or "trophic cascades" in terrestrial ecosystems was laid in a major series of exclosure experiments conducted in the Chihuahuan Desert of southeastern Arizona. Manipulation of food and the removal of one or both seed eating rodents and ants resulted in large changes in the abundance and species composition of annual plants in the enclosures. Integrity of plant communities is essential to preserving biodiversity, so these experiments raised an early warning flag to conservationists.

In many parts of North America, extirpation of dominant predators has resulted in a phenomenon known as "mesopredator release" in areas supporting small to midsized predators (foxes, skunks, raccoons, opossums, feral and domestic housecats). In such areas, mesopredators act by default as surrogate top predators. This has resulted in modified niche exploitation, altered diversity, and other ripple effects in the population structure of the community. Local elimination of coyotes, for example, allows the mesopredators to increase in number, thereby imposing added predator pressure on prey, such as quail, grouse, ducks, nightjars, and certain warblers. Mesopredator release has also been blamed for the decline or disappearance of gamebirds, songbirds, and other small vertebrates from a number of North American terrestrial ecosystems—including scrub habitats, grasslands, prairie wetlands, and eastern deciduous forests.

Extirpation of top predators has released herbivore populations in parts of the United States with consequences that are just beginning to come to light. Overbrowsing by white-tailed deer is decisively altering the pattern of tree regeneration in some eastern forests and is threatening certain endangered plants with extinction. Elsewhere in North America, introduced ungulates, especially Eurasian boar have increased to such a degree that they are destroying wildflower beds and altering tree regeneration patterns in forests. It hardly needs to be emphasized that rapid, large-scale, and unpredictable changes in forest composition represent a chilling threat to biodiversity.

For another case, let us return to Lago Guri in Venezuela, where recently created islands are experiencing cataclysmic biological change. In a predator-free environment, three generalist herbivores have each increased in abundance by more than an order of magnitude. Howler monkeys on some islands have attained densities equivalent to 500 per square kilometer whereas mainland densities are typically between 20 and 40 per square kilometer. Densities of iguanas and leaf-cutter ants have similarly exploded.

Ongoing studies of forest regeneration on these islands reveal little successful reproduction of canopy trees. Saplings die faster than they appear, so the forests on these small of herbivore-impacted islands were in sharp decline. Extraordinarily elevated densities of leaf-cutter ants were primarily to blame for high sapling mortality. In the absence of "normal" biological interactions, the remnant ecosystems of these islands have spun out of control. It seems inevitable that most of the plant and animal species that survived the initial contraction in area will be extirpated within one or two tree replacement cycles.

Vegetation change in the Lago Guri islands and in portions of the United States occupied by hyperabundant populations of white-tailed deer and Eurasian boar offer startling examples of trophic cascades—examples that mirror findings from deserts, lakes, and Pacific kelp forests. To prevent ecosystems all over North America from experiencing similar convulsions brought about by trophic cascades, the full spectrum of ecological processes that operates to perpetuate biodiversity—especially predation—must be widely maintained.

Where top predators have been extirpated and their reestablishment is impractical, can trophic cascades be avoided? Perhaps worst-case scenarios can be avoided through interventions of various sorts. But no human effort can accurately simulate the effects of real predators, because these animals have impacts on many prey species simultaneously and interact with prey populations in complex ways that are seldom understood. Nevertheless, the worst consequences of trophic cascades might be forestalled or ameliorated though the hunting of herbivores and trapping of mesopredators. The most severe impacts of overly abundant mesopredators and consumers appear in localities where predators are absent and hunting and trapping are prohibited.

Predators prevent prey populations and mesopredators from exploding into overabundance while rarely, if ever, driving prey to extinction. Prey species, such as seed dispersers, seed predators, or herbivores, are thereby regulated within definite upper and lower bounds. The operation of such feedback mechanisms can be likened to "a balance of Nature." Nature stays in balance so long as a fauna remains intact and the full suite of ecological processes operates unhindered. It is when Nature falls out of balance—when there are too many consumers and mesopredators (or not enough)—that species begin to disappear and humans begin to notice. But what humans notice is only that some favored species or another has disappeared. Hidden in the workings of a Nature we are only beginning to understand, the cause remains obscure.

ANOTHER KEY TO BIODIVERSITY

In terrestrial ecosystems, top-down and bottom-up processes operate simultaneously and are of particular importance to the functioning of natural ecosystems. This seemingly contradictory statement results not only from the complexity of food web structure but also from flexibility in the behavior of individual species—such as the tendency for prey to quickly eat in the presence of predators and the ability of plants to increase their investment in anti-herbivore defenses in response to being eaten.

Although herbivores large enough to be invulnerable to predators and herd-forming migratory ungulates tend to be regulated from the bottom up, megaherbivores concurrently exert top-down forces through their effects on vegetation. Both groups of species may have been prominent over much of the earth's surface prior to megafaunal overkill, but they have been reduced by human persecution to a tiny fraction of their former geographical occurrence. What remains nearly everywhere else are drastically truncated mammal communities that are regulated largely through top-down processes.

The evidence overwhelmingly supports the strong top-down role of large carnivores in regulating prey populations—and thereby stabilizing the trophic structure of terrestrial ecosystems. Loss of top predators results in hyperabundance of consumers playing a variety of trophic roles (herbivores, seed dispersers, seed predators) and in mesopredator release. Overabundance of consumers and mesopredators, in turn, results in trophic cascades that lead to multiple effects—including the direct elimination of plant populations from overbrowsing/grazing, reproductive failure of canopy tree species, and the loss of groundnesting birds and probably other small vertebrates.

In sum, then, our current knowledge about the natural processes that maintain biodiversity suggests a crucial and irreplaceable regulatory role of top predators. The absence of top predators appears to lead inexorably to ecosystem simplification accompanied by a rush of extinctions. Therefore, efforts to conserve North American biodiversity in interconnected megareserves will have to place a high priority on reestablishing top predators wherever they have been locally extirpated. If steps are not taken in the interim to restore the full gamut of natural abiotic and biotic processes that maintain biodiversity, efforts to halt extinction through legislated mechanisms (such as the Endangered Species Act) will be overwhelmed by irresistible biological forces. It is only by providing the conditions that allow Nature to remain in balance that biodiversity can be perpetuated over the long run.

REBUILDING AFTER COLLAPSE

JOHN DAVIS

Let's face it: we've lost, at least for this century. Despite decades of valiant efforts on the part of conservation, environmental, and peace activists the world over, the extinction crisis continues to broaden and deepen. Rates of loss in the natural world continue to accelerate at all levels, from genes to species to ecosystems. In the human realm, cultures, languages, and ways of life are being exterminated nearly as quickly. Our efforts to "save the world" may have slowed the extinction vortex, but countless life forms keep slipping away even as we stave off the latest assaults on our few charismatic champion species and communities.

After 10,000 plus years of making our living by wrecking wild Nature, through agriculture and industry, we must learn how to base our livelihoods on restoring and preserving wild Nature, and human communities compatible therewith. It is time for activists, teachers, and leaders to start plotting how to make conservation durable come what may—global climatic mayhem, pandemics, economic disintegration, war, famine, fascism, and so forth. It is also time for those of us laboring to save the world to think about how to restore the world after the industrial global growth economy finally crumbles.

Far from this being reason to surrender, the imminence of industrial collapse gives us still more reason to protect every bit of wild Nature we possibly can. My thesis is not that we in the conservation and environmental communities have failed; rather, I believe forces largely beyond our control will continue wrecking the natural world, despite our defensive efforts, until the industrial economy collapses. We need to extend our efforts, to expand and durably protect natural areas and other parts of a whole Earth through cataclysms and beyond.

Some years back, even before the full implications of the extinction crisis and global overheating were understood, deep ecologist Arne Naess presciently said he was a pessimist for the 21st century, but an optimist for the 22nd. That's a view more of us should consider. The future is unlikely to be much like the present. For a time, it may be frightfully bleak.

Climate will destabilize, rising seas will flood much of peopled and arable land, extinctions will cascade, wars over dwindling resources will erupt, plagues will douse exploding populations, and people wracked with fear will terrorize each other and turn to dictatorial demagogues as leaders to save them from a world out of control. Forgive my grim scenario. The following are the main reasons why the ruination of the natural world continues apace:

WHY WE'VE LOST

Over-Population and Consumption—We modern humans are orders of magnitude too many. With a few million *Homo sapiens* gently sprinkled across Africa 100,000 years ago, we may actually have functioned well as plain members of the biotic community, subsisting in natural ecosystems without diminishing them. With six and a half billion people rapidly multiplying and consuming more and more resources, less and less habitat is left for wild creatures. Already, we humans, one species among perhaps 100 million, are consuming nearly 40 percent of Earth's net primary productivity. If the conservation and environmental movements are partly to blame for failing to reign in industrial civilization, that blame should largely be focused on our failure to confront the fundamental problem of too many people consuming too many resources. For several centuries, national economies—and now one global economy—have been based on robbing Nature. Only an economy based on giving back to and restoring Nature could save the day now. Where are the messengers who could convince six and a half billion people of this? They are lost in the multitudes.

Technology—Another causative and self-reinforcing factor in the degradation of the natural world is mega-technology, technologies big and powerful enough to speed our exploitation of natural resources, technologies that tend to concentrate power in few and greedy hands. Industrial agriculture, in particular, has sterilized farmscapes, produced animal factories, polluted waterways, captured huge subsidies, and displaced family farmers, all at the expense of Nature and our rural communities. Scholars of technology such as Jerry Mander, Helena Norberg-Hodge, Andrew Kimbrell, Stephanie Mills, Vandana Shiva, Colin Hines, Wendell Berry, and Wolfgang Sachs have been arguing persuasively for years against the popular assumption that technology is neutral. Mander, especially, has shown that technologies predispose themselves in certain ways—too often, toward greater and more harmful exploitation of the natural world and less fortunate peoples. No critical review process is in place to carefully assess new technologies before they are unleashed on the world. Perhaps the most alarming aspect of mega-technology is how it has facilitated the concentration of information power in the wires and boardrooms of a small number of transnational corporations. Mander warns that most of the world's people now receive their news and their commentaries, and thus increasingly their views, from roughly ten huge corporations. By now, most conservationists and environmentalists would acknowledge that automobiles, televisions, and pesticides have harmed the biosphere. Yet opposition to these technologies is seldom heard, and the computerization of the world proceeds virtually unchallenged.

Short-term Priorities—Human nature predisposes us to focus on short-term rewards for ourselves and our families, not to worry about long-term or distant problems. We *Homo sapiens* are not inherently bad, not inevitably greedy, but we are creatures of habit and immediacy. Entrenched habits—such as dependence on cars, central heating, refined and globally traded foods, and electricity—are very hard to break. Well-being, for most of the world's people, has come to be perceived in terms of material affluence. It is our nature to seek security for ourselves and our families, not to be concerned about the whole biosphere. Expanding people's minds and hearts so that they see security and well-being in terms of planetary health would constitute a paradigm shift that no subset of humanity is likely to ever have the power to bring about.

Trapped in the System—Making the sorts of changes that would be necessary to reconcile our species with the rest of the biotic world would be very difficult at the individual, as well as the societal, level. Many of us have jobs that require commutes, family members needing medical attention or living far away, children clamoring for the latest toys, and avocations that make this crowded world bearable but exact a heavy toll on it. Sure, switching from gas guzzlers to hybrid cars, recycling, using energy-efficient appliances, better insulating homes, planting vegetable gardens, buying from local farms and so forth are relatively painless steps to take us in the right direction. But we have not even succeeded in persuading most Americans to take these small steps.

There is little, if any, evidence to suggest that internal combustion engines, cathode ray tubes, firearms and explosives, biocides, or large-scale industrial and agricultural production can be made compatible with the protection and restoration of biological diversity. A society at peace with the natural world would be small in numbers of people and rates of consumption and would be powered with muscles and sunshine, not motors and fossil fuels. Yet who among us is ready to renounce cars, airplanes, computers, appliances, coffee, bananas, and central heating? Who among us can, without greatly disrupting the lives of our loved ones? Bill McKibben noted years ago in *The End of Nature* that it is one thing to call for factories to clean up their emissions; it is quite another to call for these factories to be shut down.

LOOKING INWARD

To their credit, the conservation and environmental movements are soul-searching as never before. The U.S. elections of 2004 have spurred a spate of provocative papers and talks by activist leaders and scholars. Carl Pope, head of the Sierra Club, has been delivering trenchant analyses of why progressives keep losing elections. Dave Foreman, head of The Rewilding Institute, has issued a clarion call for conservationists to stand strong by their principles of protecting wild Nature for its own sake. The most controversial of recent papers is probably "The Death of Environmentalism" by Michael Schellenberger and Ted Nordhause. The "Death" paper is notable as much for what it does not say as for what it does. It does argue persuasively that environmentalists have too narrowly defined their cause and that's largely why we've been unable to deal effectively with global warming. What "Death" neglects to say is that higher fuel efficiency is not the solution to pollution: eliminating cars and factories is. Of course, saying this would virtually guarantee dismissal as politically unrealistic. How sad that destroying life on Earth is not instead viewed as politically untenable.

Some excellent new books are also challenging people to consider the fate of the Earth. E. O. Wilson's *Future of Life*, Jared Diamond's *Collapse*, Richard Wright's *Short History of Progress*, the updated *Limits to Growth: 30 Years Later*, are among the most insightful and informative in the Cassandran genre. Yet none of these nor any popular work reaches the obvious, politically unacceptable conclusion: *industrial globalization is incompatible with wild Nature.* Perhaps the closest a recent popular author has come to reaching this discomfiting conclusion is James Kunstler in *The Long Emergency.* His analysis of the perils of dependency on foreign oil and warnings about the impending breakdown of the oil economy are incisive and convincing, but his focus is on human society. He mostly leaves for others the study of what industrial collapse could mean for Nature.

WHY OUR LOSS IS NOT THE END

Our failure to save the world, however, is not the end of the world. Gaia, the Creation, Mother Earth, the dance of Life—however we wish to describe Planet Earth—will not allow one haughty member of its community to destroy all the rest. Mother Earth has been surprisingly lenient with her unruly children, but the increasingly erratic climate, stronger storms, spreading epidemics, and recurring famines foreshadow harsher treatment if *Homo sapiens* continues to misbehave. We humans will not win the war against Nature. With the more optimistic of my conservation colleagues, I take as a matter of faith Jack London's immortal words: "The wild must win in the end."

Such faith can keep us going, working tirelessly to save every bit of wild Nature we possibly can. The wild places we save today will serve as the seeds of recovery tomorrow. This is reason for securing every bit of wild Nature we possibly can, especially those pieces still of original, or primal, quality.

The resilience and renewability of life are manifest especially in those landscapes that humans have departed or allowed to recover after earlier exploitation. To cycle through the abandoned farm country of northern New England and New York, for instance, is to see Nature healing, rewilding, at a remarkable pace. Vermont has gone from 80 percent agricultural fields to 80 percent woodlands in a matter of a few human generations. The regenerating forests are not as rich and intact as the original forests, but they do retain much of the region's native diversity and they do demonstrate Nature's powers of renewal. The farms that are left are generally small and better integrated into the landscape than the huge monocultures dominating much of the United States. If humans can find the wisdom and humility to step back from large parts of the planet, wild Nature will rebound vigorously.

MAKING CONSERVATION DURABLE AND RESILIENT

I have quickly outlined what most defenders of the natural world know deep down to be true, though we dare not admit in public. The remaining points I wish to make are that—given the prospect of coming cataclysms—we Nature advocates need to be: (1) giving increased thought and action to ensuring the durable protection of natural areas and their creatures, come what may; and, (2) spreading good ideas that might help the survivors of industrial collapse rebuild in ways harmonious with Nature.

BUILDING ALLIANCES

Forming strong alliances based on shared life-affirming values should be a top priority for all who seek a better world. Michael Soulé, cofounder of the Society for Conservation Biology and the Wildlands Project and one of those rare scientists who understands people and politics, too, makes the case for a lasting alliance of what he terms the "life-affirming movements," which he identifies as humanitarianism, animalism, and naturism. Soulé explains: "Animalism is all of the animal protection movements from welfare to liberation; Humanitarianism refers to all movements dedicated to the improvement of the human condition; Naturism means nature, biodiversity, and wilderness protection." He argues persuasively that if the three main life-affirming movements join forces, we'll be formidable. We need to listen to, talk with, share stories and values, and find common ground with good people of other good causes, then to stand and defend that common ground, through the coming chaos. Speak your heart honestly, Soulé urges, and you will find friends in unexpected places.

SAVING EVERY ACRE

Meanwhile, back at the range, forest, river, and lake, we must save every acre we possibly can, employing all the tools already in our belt and others we may add as we broaden our outreach. Every acre secured constitutes at least a small victory. Although we conservationists and environmentalists probably cannot save the world, we can save significant acreage. At least some of the natural areas we secure now will survive an industrial melt-down.

STRATEGIC LANDSCAPE ACTIVISM

Given that wild Nature is being extinguished at rates far faster than we can match with our land saving, our priorities must be urgently and strategically focused.

We must preserve, expand, and reconnect all large roadless areas worldwide, and the bigger the better. We must provide safe passage through agricultural and other human use lands to these intact natural habitats. The Earth's few remaining intact ecosystems must be made high priorities for protection: the boreal forests in North America and Eurasia; rainforests in the Amazon, the Guianas, southern Chile, Central Africa, New Guinea, and British Columbia, and Alaska; grasslands of Patagonia; savannas of southern and eastern Africa; deserts of Australia, southwestern United States and northern Mexico; remote mountain ranges, including much of the Andes, Himalayas, Pamirs, Alaska Range, and Rockies; polar regions; wetlands of Iberá and Pantanal in South America, and Okavango Delta of Botswana; coral reefs of Belize, Australia, Indonesia, and Malaysia; any major estuaries still remote from commerce routes; circumpolar great lakes; and deep sea ecosystems. Top priority might fairly be given to the five great frontier forests—Canada and Alaska's boreal, Russia's boreal, Amazon and Guianan rain forest, Central African rain forest, and New Guinean rain forest—and intact coral reefs and estuaries, as the lungs and nurseries of the planet.

Provided that conservation measures are strong and durable, however, even a small preserve in a mostly humanized landscape may contribute to the ultimate rewilding of the planet. Especially important among smaller areas are the thirty or so "biological hotspots" identified by Norman Myers, Russ Mittermeir, and E. O. Wilson. Comprising but a small fraction of Earth's surface, these hotspots are thought to support nearly half the species in the world. The affluent nations can help the less developed nations of the tropics (where most of the hotspots are) afford such ambitious conservation efforts through debt forgiveness, appropriate technology support, training of rangers, and other measures. Sparsely populated or partially recovering landscapes such as the Great Plains where buffalo and wolves are slowly returning, remote parts of the Pacific Coast, the Adirondacks and Northern Appalachians of North America, and mountainous regions of Eastern Europe also merit strategic consideration.

ENLARGING AND RECONNECTING WILDLANDS

Enlarging and reconnecting existing parks and wilderness areas is a related imperative. *Ghost Bears* by Ed Grumbine, *Saving Nature's Legacy* by Reed Noss and Allen Cooperider, *Continental Conservation* edited by Michael Soulé and John Terborgh, and *Rewilding North America* by Dave Foreman eloquently explain why and how we should expand and reunite protected areas wherever possible. These books and the regional reserve designs of the Wildlands Project can serve as guides for present and future conservationists.

STRENGTHENING CONSERVATION EASEMENTS

At present, one of our most powerful tools—conservation easements—require better enforcement and higher standards. Easements that allow continued commercial exploitation or that are not strictly enforced do not constitute durable conservation. Easements should aim for the highest possible level of protection, Forever Wild, wherever possible and should be carefully monitored and adhered to on the ground. Agricultural easements should include biodiversity conservation components.

CLOSING ROADS

Politically difficult though it will be, another necessary step toward durable conservation is closing roads and other motorized access points and routes into wild areas. The surest way to keep an area Forever Wild is to forever exclude the machines. A happy irony could be in store: As the industrial global economy starts to crumble, road closures may become more feasible. Roads bring off-road vehicles, guns, traps, pollutants, alien species, poachers, and other problems. In general, the bigger and less accessible a wild area is, the more durable will be its conservation.

ACKNOWLEDGING PAST MISTAKES

We have not lost this war for lack of effort or intelligence. Countless books have been written on the ills of modern societies, and we should heed the truth of these (including works by Ed Abbey, Wendell Berry, Barbara Kingsolver, Jerry Mander, Edward Goldsmith, Simon Retallick, Bill McKibben, Stephanie Millis, Scott Russell Sanders...). We've lost the war because a few thousand—or even a few million—idealistic Earth defenders are not enough to halt, much less reverse, humanity's 10,000-year march against wild Nature. Whether or not we actually rebound from a collapse will depend on our ability to acknowledge our past mistakes.

As noted above, these mistakes include overlooking overpopulation, technological optimism, and myopia. They also include creeping resourcism.

Many conservationists have increasingly and uncritically accepted the "working landscape" model of conservation. Managed timberlands, crop, and range lands may suffice as buffers, or stepping stone habitats in some places, but the basic green infrastructure should consist of big wild cores connected by wildways, and these do not admit exploitation.

PLANTING THE SEEDS OF RECOVERY

Bolstering the resilience of wild Nature depends on identifying, mapping, and protecting the green infrastructure of a rewilding world. This should be the first big outward task of the survivors of industrial collapse, if it has not already been done by then. Another part of the process of completing greenprints for each region is calculating just how many people a region can support, while still meeting the needs of all the native denizens. We should commence these calculations now, while we have use of computers and satellites and geographic information systems. Better, we current-day conservationists will complete the task before the crash, though a rapidly overheating climate may force modifications in plans.

Of course, a better, wilder future world will only be possible and sustainable if good stories and sound ideas and generous values have been spread and adopted. These good stories and ideas and values, which we present-day people need to be planting now, would inculcate in survivors and rebuilders a strong awareness of limits—particularly in terms of population, technology, and consumption. These limits can be at least roughly calculated today, using information technologies that may no longer be around after economic or social collapse. A basic value that infuses much old mythology, is arrogantly scorned in much modern media, and needs to be reincorporated in our new stories and myths, is the virtue of self-restraint, as eloquently advocated through decades of writing by the farmer and poet Wendell Berry.

To plant these seeds of future recovery, a new genre of literature and other arts and sciences is needed to realistically yet idealistically describe and depict what paths post-petroleum peoples might best follow. Yes, some fine ecotopian novels have been written, and plenty of works on the perils of today's dominant industrial global economy; but the field of prophecy is wide open. Far from complete is the exploration needed to determine what forms of social organization, subsistence, economy, governance, law, custom, and habitation are truly ecologically just and sustainable. Dan Imhoff's recent book, *Farming with the Wild,* is a shining example of the sorts of new exploration and explanation needed. The artistic and other media needed to communicate knowledge of sustainable infrastructure and livelihoods lastingly into the future constitute new territory for most people in the life-affirming movements. In short, though we advocates for the natural world are already overworked, we need to add two areas of focus to our agendas: *durability* and *resilience*—helping Nature survive the coming chaos and spreading the good ideas that will allow human communities to rebuild more sensibly after collapse, even while Nature is rebuilding the green infrastructure.

PERSONAL ACKNOWLEDGMENTS

My generation, impressionable and hopeful when the first Earth Day brought environmentalism into the mainstream, has proven no more resistant to the trappings of industrial civilization than our parents. Nonetheless, even if we cannot abide by them ourselves, instilling in our children good, ecological, generous values will bear fruit, for the extinction crisis will soon be impossible to ignore, and the ecologically minded and informed youth of the late industrial world may beget the leaders of a rewilding world. With a little more help from us voices for the downtrodden, wild Earth can survive—albeit, in a badly wounded state—humanity's sacrilegious war against other life forms and then begin rebuilding, re-evolving, with a much smaller number of people, who will be wiser and more generous if we alive today have done our work.

In the end, then, we should do all the good work and none of the needless consumption that we've been doing. Let's add to that careful long-term planning to increase the odds that wild places and creatures and processes will stay forever wild and to ensure that good ideas and values and visions will likewise carry on through the crashing seas ahead.

I do not expect to see the world whole and healed. But I do retain hope that my nephews' children or grandchildren will be fortunate enough to see the healing well underway. To them, and others of the future, I leave this urgent plea: *Be fair and generous to all life forms. Carry your own weight and that of your elders. Keep your numbers low, your technology simple, and let the world be wild!*

Section IV

SOCIETY & CULTURE

If food and farming remain the most profound connection with Mother Earth experienced by a majority of people, surely those bonds are being stretched thin. Understanding and sleuthing the complexities and startling revelations of the modern food chain have become the detective work of a new generation of writers, journalists, nutritionists, and ecologists. This is no intellectual exercise. Food and farming represent one of the core issues of our time. If we can't develop systems and values to produce foods without destroying the land and threatening our health, how will we possibly confront larger looming challenges?

Innovative farming practices and marketing arrangements are reconnecting us to the land. For decades, organic farming was belittled as a niche activity that would cause mass starvation and encroach upon wildlands if adopted by the world's "real farmers." History is proving otherwise, as the movement evolves and new information on natural fertility practices and yields come to light. Reaching beyond organic certification, the market is embracing watershed stewardship, biodiversity protection, and other holistic approaches. Though progress has been slow, emerging eco-labels have hard-won lessons to share. Creative marketing and a redirection of federal subsidies may ultimately help those mid-sized farmers that are falling by the wayside because they can't compete in global or local markets. But it will take a hunger for a new diet that centers around locally grown, home-cooked meals and a willingness to pay more for what we eat. In the end, a concerted effort among citizens, farmers, chefs, government agencies, conservationists, health officials, and others will no doubt be required to make food production and farming an act that honors the very land we depend on for nourishment.

THE WAY WE LIVE NOW

MICHAEL POLLAN

Sometimes even complicated social problems turn out to be simpler than they look. Take America's "obesity epidemic," arguably the most serious public-health problem facing the country. Three of every five Americans are now overweight, and some researchers predict that today's children will be the first generation of Americans whose life expectancy will actually be shorter than that of their parents. The culprit, they say, is the health problems associated with obesity.

You hear several explanations. Big food companies are pushing supersize portions of unhealthful foods on us and our children. We have devolved into a torpid nation of couch potatoes. The family dinner has succumbed to the fast-food outlet. All these explanations are true, as far as they go. But it pays to go a little further, to look for the cause behind the causes. Which, very simply, is this: when food is abundant and cheap, people will eat more of it and get fat. Since 1977, an American's average daily intake of calories has jumped by more than 10 percent. Those 200 or so extra calories have to go somewhere. But the interesting question is, Where, exactly, did all those extra calories come from in the first place? And the answer takes us back to the source of all calories: the farm.

It turns out that we have been here before, sort of, though the last great American binge involved not food, but alcohol. It came during the first decades of the 19th century, when Americans suddenly began drinking more than they ever had before or have since, going on a collective bender that confronted the young republic with its first major public-health crisis—the obesity epidemic of its day. Corn whiskey, suddenly superabundant and cheap, was the drink of choice, and in the 1820s the typical American man was putting away half a pint of the stuff every day. That works out to more than five gallons of spirits a year for every American. The figure today is less than a gallon.

As W. J. Rorabaugh tells the story in "The Alcoholic Republic," we drank the hard stuff at breakfast, lunch, and dinner, before work and after and very often during. Employers were expected to supply spirits over the course of the workday; in fact, the modern coffee break began as a late-morning whiskey break called "the elevenses." (Just to pronounce it makes you sound tipsy.) Except for a brief respite Sunday mornings in church, Americans simply did not gather—whether for a barn raising or quilting bee, corn husking or political campaign—without passing the jug. Visitors from Europe—hardly models of sobriety themselves—marveled at the free flow of American spirits. "Come on then, if you love toping," the journalist William Cobbett wrote his fellow Englishmen in a dispatch from America. "For here you may drink yourself blind at the price of sixpence."

The results of all this toping were entirely predictable: a rising tide of public drunkenness, violence and family abandonment, and a spike in alcohol-related diseases. Several of the founding fathers—including George Washington, Thomas Jefferson, and John Adams—denounced the excesses of the "alcoholic republic," inaugurating the American quarrel over drinking that would culminate a century later in Prohibition.

But the outcome of our national drinking binge is not nearly as relevant to our present predicament as its underlying cause. Which, put simply, was this: American farmers were producing way too much corn, especially in the newly settled areas west of the Appalachians, where fertile soil yielded one bumper crop after another. Much as it has today, the astounding productivity of American farmers proved to be their own worst enemy, as well as a threat to the public health. For when yields rise, the market is flooded with grain, and its price collapses. As a result, there is a surfeit of cheap calories that clever marketers sooner or later will figure out a way to induce us to consume.

In those days, the easiest thing to do with all that grain was to distill it. The Appalachian range made it difficult and expensive to transport surplus corn from the lightly settled Ohio River Valley to the more populous markets of the East, so farmers turned their corn into whiskey—a more compact and portable "value-added commodity." In time, the price of whiskey plummeted, to the point that people could afford to drink it by the pint, which is precisely what they did.

Nowadays, for somewhat different reasons, corn (along with most other agricultural commodities) is again abundant and cheap, and once again the easiest thing to do with the surplus is to turn it into more compact and portable value-added commodities: corn sweeteners, corn-fed meat and chicken, and highly processed foods of every description. The Alcoholic Republic has given way to the Republic of Fat, but in both cases, before the clever marketing, before the change in lifestyle, stands a veritable mountain of cheap grain. Until we somehow deal with this surfeit of calories coming off the farm, it is unlikely that even the most well-intentioned food companies or public-health campaigns will have much success changing the way we eat.

The underlying problem is agricultural overproduction, and that problem (while it understandably never receives quite as much attention as underproduction) is almost as old as agriculture itself. Even in the Old Testament, there's talk about how to deal not only with the lean times but also with the fat: the Bible advises creation of a grain reserve to smooth out the swings of the market in food. The nature of farming has always made it difficult to synchronize supply and demand. For one thing, there are the vagaries of nature: farmers may decide how many acres they will plant, but precisely how much food they produce in any year is beyond their control.

The rules of classical economics just don't seem to operate very well on the farm. When prices fall, for example, it would make sense for farmers to cut back on production, shrinking the supply of food to drive up its price. But in reality, farmers do precisely the opposite, planting and harvesting more food to keep their total income from falling, a practice that of course depresses prices even further. What's rational for the individual farmer is disastrous for farmers as a group. Add to this logic the constant stream of improvements in agricultural technology (mechanization, hybrid seed, agrochemicals, and now genetically modified crops—innovations all eagerly seized on by farmers hoping to stay one step ahead of falling prices by boosting yield), and you have a sure-fire recipe for overproduction—another word for way too much food.

All this would be bad enough if the government weren't doing its best to make matters even worse, by recklessly encouraging farmers to produce even more unneeded food. Absurdly, while one hand of the federal government is campaigning against the epidemic of obesity, the other hand is actually subsidizing it, by writing farmers a check for every bushel of corn they can grow. We have been hearing a lot lately about how our agricultural policy is undermining our foreign-policy goals, forcing third-world farmers to compete against a flood tide of cheap American grain. Well, those same policies are also undermining our public-health goals by loosing a tide of cheap calories at home.

While it is true that our farm policies are making a bad situation worse, adding mightily to the great mountain of grain, this hasn't always been the case with government support of farmers, and needn't be the case even now. For not all support programs are created equal, a fact that has been conveniently overlooked in the new free-market campaign to eliminate them.

In fact, farm programs in America were originally created as a way to shrink the great mountain of grain, and for many years they helped to do just that. The Roosevelt administration established the nation's first program of farm support during the Depression, though not, as many people seem to think, to feed a hungry nation. Then, as now, the problem was too much food, not too little; New Deal farm policy was designed to help farmers reeling from a farm depression caused by what usually causes a farm depression: collapsing prices due to overproduction. In Churdan, Iowa, recently, a corn farmer named George Naylor told me about the winter day in 1933 his father brought a load of corn to the grain elevator, where "the price had been 10 cents a bushel the day before," and was told that suddenly, "the elevator wasn't buying at any price." The price of corn had fallen to zero.

New Deal farm policy, quite unlike our own, set out to solve the problem of overproduction. It established a system of price supports, backed by a grain reserve, that worked to keep surplus grain off the market, thereby breaking the vicious cycle in which farmers have to produce more every year to stay even.

It is worth recalling how this system worked, since it suggests one possible path out of the current subsidy morass. Basically, the federal government set and supported a target price (based on the actual cost of production) for storable commodities like corn. When the market price dropped below the target, a farmer was given an option: rather than sell his harvest at the low price, he could take out what was called a "nonrecourse loan," using his corn as collateral, for the full value of his crop. The farmer then stored his corn until the market improved, at which point he sold it and used the proceeds to repay the loan. If the market failed to improve that year, the farmer could discharge his debt simply by handing his corn over to the government, which would add it to something called, rather quaintly, the "ever-normal granary." This was a grain reserve managed by the USDA, which would sell from it whenever prices spiked (during a bad harvest, say), thereby smoothing out the vicissitudes of the market and keeping the cost of food more or less steady—or "ever normal."

This wasn't a perfect system by any means, but it did keep cheap grain from flooding the market and by doing so supported the prices farmers received. And it did this at a remarkably small cost to the government, since most of the loans were repaid. Even when they weren't, and the government was left holding the bag (i.e., all those bushels of collateral grain), the USDA was eventually able to unload it, and often did so at a profit. The program actually made money in good years. Compare that with the current subsidy regime, which costs American taxpayers about $19 billion a year and does virtually nothing to control production.

So why did we ever abandon this comparatively sane sort of farm policy? Politics, in a word. The shift from an agricultural-support system designed to discourage overproduction to one that encourages it dates to the early 1970s—to the last time food prices in America climbed high enough to generate significant political heat. That happened after news of Nixon's 1972 grain deal with the Soviet Union broke, a disclosure that coincided with a spell of bad weather in the farm belt. Commodity prices soared, and before long so did supermarket prices for meat, milk, bread, and other staple foods tied to the cost of grain. Angry consumers took to the streets to protest food prices and staged a nationwide meat boycott to protest the high cost of hamburger, that American birthright. Recognizing the political peril, Nixon ordered his secretary of agriculture, Earl (Rusty) Butz, to do whatever was necessary to drive down the price of food.

Butz implored America's farmers to plant their fields "fence row to fence row" and set about dismantling 40 years of farm policy designed to prevent overproduction. He shuttered the ever-normal granary, dropped the target price for grain, and inaugurated a new subsidy system, which eventually replaced nonrecourse loans with direct payments to farmers. The distinction may sound technical, but in effect it was revolutionary. For instead of lending farmers money so they could keep their grain off the market, the government offered to simply cut them a check, freeing them to dump their harvests on the market no matter what the price.

The new system achieved exactly what it was intended to: the price of food hasn't been a political problem for the government since the Nixon era. Commodity prices have steadily declined, and in the perverse logic of agricultural economics, production has increased, as farmers struggle to stay solvent. As you can imagine, the shift from supporting agricultural prices to subsidizing much lower prices has been a boon to agribusiness companies because it slashes the cost of their raw materials. That's why Big Food, working with the farm-state congressional delegations it lavishly supports, consistently lobbies to maintain a farm policy geared to high production and cheap grain. (It doesn't hurt that those lightly populated farm states exert a disproportionate influence in Washington, since it takes far fewer votes to elect a senator in Kansas than in California. That means agribusiness can presumably "buy" a senator from one of these underpopulated states for a fraction of what a big-state senator costs.)

But as we're beginning to recognize, our cheap-food farm policy comes at a high price: first there's the $19 billion a year the government pays to keep the whole system afloat; then there's the economic misery that the dumping of cheap American grain inflicts on farmers in the developing world; and finally there's the obesity epidemic at home—which most researchers date to the mid-70s, just when we switched to a farm policy consecrated to the overproduction of grain. Since that time, farmers in the United States have managed to produce 500 additional calories per person every day; each of us is, heroically, managing to pack away about 200 of those extra calories per day. Presumably the other 300—most of them in the form of surplus corn—get dumped on overseas markets or turned into ethanol.

Cheap corn, the dubious legacy of Earl Butz, is truly the building block of the "fast-food nation." Cheap corn, transformed into high-fructose corn syrup, is what allowed Coca-Cola to move from the svelte 8-ounce bottle of soda ubiquitous in the 70s to the chubby 20-ounce bottle of today. Cheap corn, transformed into cheap beef, is what allowed McDonald's to supersize its burgers and still sell many of them for no more than a dollar. Cheap corn gave us a whole raft of new highly processed foods, including the world-beating chicken nugget, which, if you study its ingredients, you discover is really a most ingenious transubstantiation of corn, from the cornfed chicken it contains to the bulking and binding agents that hold it together.

You would have thought that lower commodity prices would represent a boon to consumers, but it doesn't work out that way, not unless you believe a 32-ounce Big Gulp is a great deal. When the raw materials for food become so abundant and cheap, the clever strategy for a food company is not necessarily to lower prices—to do that would only lower its revenues. It makes much more sense to compete for the consumer's dollar by increasing portion sizes—and as Greg Critser points out in his recent book *Fat Land,* the bigger the portion, the more food people will eat. So McDonald's tempts us by taking a 600-calorie meal and jacking it up to 1,550 calories. Compared with that of the marketing, packaging and labor, the cost of the added ingredients is trivial.

Such cheap raw materials also argue for devising more and more highly processed food, because the real money will never be in selling cheap corn (or soybeans or rice) but in "adding value" to that commodity. Which is one reason that in the years since the nation moved to a cheap-food farm policy, the number and variety of new snack foods in the supermarket have ballooned. The game is in figuring out how to transform a penny's worth of corn and additives into a $3 bag of ginkgo biloba-fortified brain-function-enhancing puffs, or a dime's worth of milk and sweeteners into Swerve, a sugary new "milk based" soft drink to be sold in schools. It's no coincidence that Big Food has suddenly "discovered" how to turn milk into junk food: the government recently made deep cuts in the dairy-farm program, and as a result milk is nearly as cheap a raw material as water.

As public concern over obesity mounts, the focus of political pressure has settled on the food industry and its marketing strategies—supersizing portions, selling junk food to children, lacing products with transfats and sugars. Certainly Big Food bears some measure of responsibility for our national eating disorder—a reality that a growing number of food companies have publicly accepted. In recent months, Kraft, McDonalds, and Coca-Cola have vowed to change marketing strategies and even recipes in an effort to help combat obesity and, no doubt, ward off the coming tide of litigation.

There is an understandable reluctance to let Big Food off the hook. Yet by devising ever-more ingenious ways to induce us to consume the surplus calories our farmers are producing, the food industry is only playing by a set of rules written by our government. (And maintained, it is true, with the industry's political muscle.) The political challenge now is to rewrite those rules, to develop a new set of agricultural policies that don't subsidize overproduction—and overeating. For unless we somehow deal with the mountain of cheap grain that makes the Happy Meal and the Double Stuf Oreo such "bargains," the calories are guaranteed to keep coming.

CAN ORGANIC FARMING FEED US ALL?

BRIAN HALWEIL

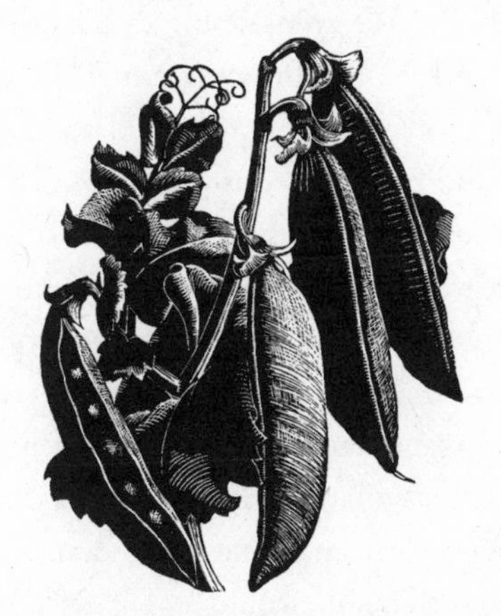

The only people who think organic farming can feed the world are delusional hippies, hysterical moms, and self-righteous organic farmers. Right?

Actually, no. A fair number of agribusiness executives, agricultural and ecological scientists, and international agriculture experts believe that a large-scale shift to organic farming would not only *increase* the world's food supply, but might be the only way to eradicate hunger.

This probably comes as a surprise. After all, organic farmers scorn the pesticides, synthetic fertilizers, and other tools that have become synonymous with high-yield agriculture. Instead, organic farmers depend on raising animals for manure, growing beans, clover, or other nitrogen-fixing legumes, or making compost and other sources of fertilizer that cannot be manufactured in a chemical plant but are instead grown—which consumes land, water, and other resources. (In contrast, producing synthetic fertilizers consumes massive amounts of petroleum.) Since organic farmers can't use synthetic pesticides, one can imagine that their fields suffer from a scourge of crop-munching bugs, fruit-rotting blights, and plant-choking weeds. And because organic farmers depend on rotating crops to help control pest problems, the same field won't grow corn or wheat or some other staple as often.

As a result, the argument goes, a world dependent on organic farming would have to farm more land than it does today—even if it means less pollution, fewer abused farm animals, and fewer carcinogenic residues on our vegetables. "We aren't going to feed 6 billion people with organic fertilizer," said Nobel Prize–winning plant breeder Norman Borlaug at a 2002 conference. "If we tried to do it, we would level most of our forest and many of those lands would be productive only for a short period of time." Cambridge chemist John Emsley put it more bluntly: "The greatest catastrophe that the human race could face this century is not global warming but a global conversion to 'organic farming'—an estimated 2 billion people would perish."

In recent years, organic farming has attracted new scrutiny, not just from critics who fear that a large-scale shift in its direction would cause billions to starve, but also from farmers and development agencies who actually suspect that such a shift could *better* satisfy hungry populations. Unfortunately, no one had ever systematically analyzed whether in fact a widespread shift to organic farming would run up against a shortage of nutrients and a lack of yields—until recently. The results are striking.

HIGH-TECH, LOW-IMPACT

There are actually myriad studies from around the world showing that organic farms can produce about as much, and in some settings much more, than conventional farms. Where there is a yield gap, it tends to be widest in wealthy nations, where farmers use copious amounts of synthetic fertilizers and pesticides in a perennial attempt to maximize yields. It is true that farmers converting to organic production often encounter lower yields in the first few years, as the soil and surrounding biodiversity recover from years of assault with chemicals. And it may take several seasons for farmers to refine the new approach.

But the long-standing argument that organic farming would yield just one-third or one-half of conventional farming was based on biased assumptions and lack of data. For example, the often-cited statistic that switching to organic farming in the United States would only yield one-quarter of the food currently produced there is based on a U.S. Department of Agriculture study showing that all the manure in the United States could only meet one-quarter of the nation's fertilizer needs—even though organic farmers depend on much more than just manure.

More up-to-date research refutes these arguments. For example, a recent study by scientists at the Research Institute for Organic Agriculture in Switzerland showed that organic farms were only 20 percent less productive than conventional plots over a 21-year period. Looking at more than 200 studies in North America and Europe, Per Pinstrup Andersen (a Cornell professor and winner of the World Food Prize) and colleagues recently concluded that organic yields were about 80 percent of conventional yields. And many studies show an even narrower gap. Reviewing 154 growing seasons' worth of data on various crops grown on rain-fed and irrigated land in the United States, University of California–Davis agricultural scientist Bill Liebhardt found that organic corn yields were 94 percent of conventional yields, organic wheat yields were 97 percent, and organic soybean yields were 94 percent. Organic tomatoes showed no yield difference.

More important, in the world's poorer nations where most of the world's hungry live, the yield gaps completely disappear. University of Essex researchers Jules Pretty and Rachel Hine looked at over 200 agricultural projects in the developing world that converted to organic and ecological approaches, and found that for all the projects—involving 9 million farms on nearly 30 million hectares—yields increased an average of 93 percent. A seven-year study from Maikaal District in central India involving 1,000 farmers cultivating 3,200 hectares found that average yields for cotton, wheat, chili, and soy were as much as 20 percent higher on the organic farms than on nearby conventionally managed ones. Farmers and agricultural scientists attributed the higher yields in this dry region to the emphasis on cover crops, compost, manure, and other practices that increased organic matter (which helps retain water) in the soils. A study from Kenya found that while organic farmers in "high-potential areas" (those with above-average rainfall and high soil quality) had lower maize yields than nonorganic farmers, organic farmers in areas with poorer resource endowments consistently outyielded conventional growers. (In both regions, organic farmers had higher net profits, return on capital, and return on labor.)

Contrary to critics who jibe that it's going back to farming like our grandfathers did or that most of Africa already farms organically and it can't do the job, organic farming is a sophisticated combination of old wisdom and modern ecological innovations that help harness the yield-boosting effects of nutrient cycles, beneficial insects, and crop synergies. It's heavily dependent on technology—just not the technology that comes out of a chemical plant.

HIGH-CALORIE FARMS

So could we make do without the chemical plants? Inspired by a field trip to a nearby organic farm where the farmer reported that he raised an amazing 27 tons of vegetables on six-tenths of a hectare in a relatively short growing season, a team of scientists from the University of Michigan tried to estimate how much food could be raised following a global shift to organic farming. The team combed through the literature for any and all studies comparing crop yields on organic farms with those on nonorganic farms. Based on 293 examples, they came up with global dataset of yield ratios for the world's major crops for the developed and the developing world. As expected, organic farming yielded less than conventional farming in the developed world for most food categories, while studies from the developing world showed organic farming boosting yields. The team then ran two models. The first was conservative in the sense that it applied the yield ratio for the developed world to the entire planet, that is, they assumed that every farm regardless of location would get only the lower-developed-country yields. The second applied the yield ratio for the developed world to wealthy nations and the yield ratio for the developing world to those countries.

"We were all surprised by what we found," said Catherine Badgley, a Michigan paleoecologist who was one of the lead researchers. The first model yielded 2,641 kilocalories ("calories") per person per day, just under the world's current production of 2,786 calories but significantly higher than the average caloric requirement for a healthy person of between 2,200 and 2,500. The second model yielded 4,381 calories per person per day, 75 percent greater than current availability—and a quantity that could theoretically sustain a much larger human population than is currently supported on the world's farmland. (It also laid to rest another concern about organic agriculture; see sidebar, "Enough Nitrogen to Go Around?")

The Michigan results imply that no additional land area is required to obtain enough biologically available nitrogen, even without including the potential for intercropping (several crops grown in the same field at the same time), rotation of livestock with annual crops, and inoculation of soil with Azobacter, Azospirillum, and other free-living nitrogen-fixing bacteria.

The team's interest in this subject was partly inspired by the concern that a large-scale shift to organic farming would require clearing additional wild areas to compensate for lower yields—an obvious worry for scientists like Badgley, who studies present and past biodiversity. The only problem with the argument, she said, is that much of the world's biodiversity exists in close proximity to farmland, and that's not likely to change anytime soon. "If we simply try to maintain biodiversity in islands around the world, we will lose most of it," she said. "It's very important to make areas between those islands friendly to biodiversity. The idea of those areas being pesticide-drenched fields is just going to be a disaster for biodiversity, especially in the tropics. The world would be able to sustain high levels of biodiversity much better if we could change agriculture on a large scale."

Badgley's team went out of the way to make its assumptions as conservative as possible: most of the studies they used looked at the yields of a single crop, even though many organic farms grow more than one crop in a field at the same time, yielding more total food even if the yield of any given crop may be lower. Skeptics may doubt the team's conclusions—as ecologists, they are likely to be sympathetic to organic farming—but a second recent study of the potential of a global shift to organic farming, led by Niels Halberg of the Danish Institute of Agricultural Sciences, came to very similar conclusions, even though the authors were economists, agronomists, and international development experts.

Enough Nitrogen to Go Around?
In addition to looking at raw yields, the University of Michigan scientists also examined the common concern that there aren't enough available sources of non-synthetic nitrogen—compost, manure, and plant residues—in the world to support large-scale organic farming. For instance, in his book *Enriching the Earth: Fritz Haber, Carl Bosch, and the Transformation of World Food Production,* Vaclav Smil argues that roughly two-thirds of the world's food harvest depends on the Haber-Bosch process, the technique developed in the early 20th century to synthesize ammonia fertilizer from fossil fuels. (Smil admits that he largely ignored the contribution of nitrogen-fixing crops and assumed that some of them, like soybeans, are net users of nitrogen, although he himself points out that on average half of all the fertilizer applied globally is wasted and not taken up by plants.) Most critics of organic farming as a means to feed the world focus on how much manure—and how much related pastureland and how many head of livestock—would be needed to fertilize the world's organic farms. "The issue of nitrogen is different in different regions," says Don Lotter, an agricultural consultant who has published widely on organic

Like the Michigan team, Halberg's group made an assumption for the differences in yields for organic farming for a range of crops and then plugged these numbers into a model developed by the World Bank's International Food Policy Research Institute (IFPRI). This model is considered the definitive algorithm for predicting food output, farm income, and the number of hungry people throughout the world. Given the growing interest in organic farming among consumers, government officials, and agricultural scientists, the researchers wanted to assess whether a large-scale conversion to organic farming in Europe and North America (the world's primary food exporting regions) would reduce yields, increase world food prices, or worsen hunger in poorer nations that depend on imports, particularly those people living in the third world's swelling megacities. Although the farming and nutrient requirements. "But lots more nitrogen comes in as green manure than animal manure."

Looking at 77 studies from the temperate areas and tropics, the Michigan team found that greater use of nitrogen-fixing crops in the world's major agricultural regions could result in 58 million metric tons more nitrogen than the amount of synthetic nitrogen currently used every year. Research at the Rodale Institute in Pennsylvania showed that red clover used as a winter cover in an oat/wheat–corn–soy rotation, with no additional fertilizer inputs, achieved yields comparable to those in conventional control fields. Even in arid and semi-arid tropical regions like East Africa, where water availability is limited between periods of crop production, drought-resistant green manures such as pigeon peas or groundnuts could be used to fix nitrogen. In Washington state, organic wheat growers have matched their nonorganic neighbor's wheat yields using the same field pea rotation for nitrogen. In Kenya, farmers using leguminous tree crops have doubled or tripled corn yields as well as suppressing certain stubborn weeds and generating additional animal fodder.

group found that total food production declined in Europe and North America, the model didn't show a substantial impact on world food prices. And because the model assumed, like the Michigan study, that organic farming would boost yields in Africa, Asia, and Latin America, the most optimistic scenario even had hunger-plagued sub-Saharan Africa exporting food surpluses.

"Modern non-certified organic farming is a potentially sustainable approach to agricultural development in areas with low yields due to poor access to inputs or low yield potential because it involves lower economic risk than comparative interventions based on purchased inputs and may increase farm level resilience against climatic fluctuations," Halberg's team concluded. In other words, studies from the field show that the yield increases from shifting to organic farming are highest and most consistent in exactly those poor, dry, remote areas where hunger is most severe. "Organic agriculture could be an important part of increased food security in sub-Saharan Africa," says Halberg.

That is, if other problems can be overcome. "A lot of research is to try to kill prejudices," Halberg says—like the notion that organic farming is only a luxury, and one that poorer nations cannot afford—"I'd like to kill this once and for all. The two sides are simply too far from each other and they ignore the realities of the global food system." Even if a shift toward organic farming boosted yields in hungry African and Asian nations, the model found that nearly a billion people remained hungry, because any surpluses were simply exported to areas that could best afford it.

WRONG QUESTION?

These conclusions about yields won't come as a surprise to many organic farmers. They have seen with their own eyes and felt with their own hands how productive they can be. But some supporters of organic farming shy away from even asking whether it can feed the world, simply because they don't think it's the most useful question. There is good reason to believe that a global conversion to organic farming would not proceed as seamlessly as plugging some yield ratios into a spreadsheet.

To begin with, organic farming isn't as easy as farming with chemicals. Instead of choosing a pesticide to prevent a pest outbreak, for example, a particular organic farmer might consider altering his crop rotation, planting a crop that will repel the pest or one that will attract its predators—decisions that require some experimentation and long-term planning. Moreover, the IFPRI study suggested that a large-scale conversion to organic farming might require that most dairy and beef production eventually "be better integrated in cereal and other cash crop rotations" to optimize use of the manure. Bringing cows back to one or two farms to build up soil fertility may seem like a no-brainer, but doing it wholesale would be a challenge—and dumping ammonia on depleted soils still makes for a quicker fix.

Again, these are just theoretical assumptions, since a global shift to organic farming could take decades. But farmers are ingenious and industrious people, and they tend to cope with whatever problems are at hand. Eliminate nitrogen fertilizer and many farmers will probably graze cows on their fields to compensate. Eliminate fungicides and farmers will look for fungus-resistant crop varieties. As more and more farmers begin to farm organically, everyone will get better at it. Agricultural research centers and universities and agriculture ministries will throw their resources into this type of farming—a far cry from their current neglect of organic agriculture, which partly stems from the assumption that organic farmers will never play a major role in the global food supply.

So the problems of adopting organic techniques do not seem insurmountable. But those problems may not deserve most of our attention; even if a mass conversion over, say, the next two decades, dramatically increased food production, there's little guarantee that would eradicate hunger. The global food system can be a complex and unpredictable beast. It's hard to anticipate how China's rise as a major importer of soybeans for its feedlots, for instance, might affect food supplies elsewhere. (It's likely to drive up food prices.) Or how elimination of agricultural subsidies in wealthy nations might affect poorer countries. (It's likely to boost farm incomes and reduce hunger.) And would less meat eating around the world free up food for the hungry? (It would, but could the hungry afford it?) In other words, "Can organic farming feed the world?" is probably not even the right question, since feeding the world depends more on politics and economics than any technological innovations.

"'Can organic farming feed the world' is indeed a bogus question," says Gene Kahn, a long-time organic farmer who founded Cascadian Farms organic foods and is now vice president of sustainable development for General Mills. "The real question is, can we feed the world? Period. Can we fix the disparities in human nutrition?" Kahn notes that the marginal difference in today's organic yields and the yields of conventional agriculture wouldn't matter if food surpluses were redistributed.

But organic farming will yield other benefits that are too numerous to name. Studies have shown, for example, that the "external" costs of organic farming—erosion, chemical pollution to drinking water, death of birds and other wildlife—are just one-third those of conventional farming. Surveys from every continent show that organic farms support many more species of birds, wild plants, insects, and other wildlife than

Food versus Fuel

Sometimes, when humans try to solve one problem, they end up creating another. The global food supply is already under serious strain: More than 800 million people go hungry every day, the world's population continues to expand, and a growing number of people in the developing world are changing to a more Western, meat-intensive diet that requires more grain and water per calorie than traditional diets do. Now comes another potential stressor: Concern about climate change means that more nations are interested in converting crops into biofuels as an alternative to fossil fuels. But could this transition remove land from food production and further intensify problems of world hunger?

For several reasons, some analysts say no, at least not in the near future. First, they emphasize that nearly 40 percent of global cereal crops are fed to livestock, not humans, and that global prices of grains and oil seeds do not always affect the cost of food for the hungry, who generally cannot participate in formal markets anyway.

Second, at least to date, hunger has been due primarily to inadequate income and distribution rather than absolute food scarcity. In this regard, a biofuels economy may actually help to reduce hunger and poverty. A recent UN Food and Agriculture Organization report

conventional farms. And tests by several governments have shown that organic foods carry just a tiny fraction of the pesticide residues of the nonorganic alternatives, while completely banning growth hormones, antibiotics, and many additives allowed in many conventional foods. There is even some evidence that crops grown organically have considerably higher levels of health-promoting antioxidants.

There are social benefits as well. Because organic farming doesn't depend on expensive inputs, it might help shift the balance toward smaller farmers in hungry nations. A 2002 report from the UN Food and Agricultural Organization noted that "organic systems can double or triple the productivity of traditional systems" in developing nations but suggested that yield comparisons offer a "limited, narrow, and often misleading picture" since farmers in these countries often adopt organic argued that increased use of biofuels could diversify agricultural and forestry activities, attract investment in new small and medium-sized enterprises, and increase investment in agricultural production, thereby increasing the incomes of the world's poorest people.

Third, biofuel refineries in the future will depend less on food crops and increasingly on organic wastes and residues. Producing biofuels from corn stalks, rice hulls, sawdust, or waste paper is unlikely to affect food production directly. And there are drought-resistant grasses, fast-growing trees, and other energy crops that will grow on marginal lands unsuitable for raising food.

Nonetheless, with growing human appetites for both food and fuel, biofuels' long-run potential may be limited by the priority given to food production if bioenergy systems are not harmonized with food systems. The most optimistic assessments of the long-term potential of biofuels have assumed that agricultural yields will continue to improve and that world population growth and food consumption will stabilize. But the assumption about population may prove to be wrong. And yields, organic or otherwise, may not improve enough if agriculture in the future is threatened by declining water tables or poor soil maintenance.

farming techniques to save water, save money, and reduce the variability of yields in extreme conditions. A more recent study by the International Fund for Agricultural Development found that the higher labor requirements often mean that "organic agriculture can prove particularly effective in bringing redistribution of resources in areas where the labour force is underemployed. This can help contribute to rural stability."

MIDDLE EARTH

These benefits will come even without a complete conversion to organic agriculture. In fact, some experts think that a more pracical way forward is a sort of middle ground, where more and more farmers adopt the principles of organic farming even if they don't follow the approach entirely. In this scenario, both poor farmers and the environment come out way ahead. "Organic agriculture is *not* going to do the trick," says Roland Bunch, an agricultural extensionist who has worked for decades in Africa and the Americas and is now with COSECHA (Association of Consultants for a Sustainable, Ecological, and People-Centered Agriculture) in Honduras. Bunch knows first-hand that organic agriculture can produce more than conventional farming among poorer farmers. But he also knows that these farmers cannot get the premium prices paid for organic produce elsewhere, and that they are often unable, and unwilling, to shoulder some of the costs and risks associated with going completely organic.

Instead, Bunch points to "a middle path," of eco-agriculture, or low-input agriculture that uses many of the principles of organic farming and depends on just a small fraction of the chemicals. "These systems can immediately produce two or three times what smallholder farmers are presently producing," Bunch says. "And furthermore, it is attractive to smallholder farmers because it is less costly per unit produced." In addition to the immediate gains in food production, Bunch suggests that the benefits for the environment of this middle path will be far greater than going "totally organic," because "something like five to ten times as many smallholder farmers will adopt it per unit of extension and training expense, because it behooves them economically. They aren't taking food out of their kids' mouths. If five farmers eliminate half their use of chemicals, the effect on the environment will be 2.5 times as great as if one farmer goes totally organic."

And farmers who focus on building their soils, increasing biodiversity, or bringing livestock into their rotation aren't precluded from occasionally turning to synthetic nitrogen or other yield-enhancing innovations in the future, particularly in places where the soils are heavily depleted. "In the end, if we do things right, we'll build a lot of organic into conventional systems," says Don Lotter, the agricultural consultant. Like Bunch, Lotter notes that such an "integrated" approach often out-performs both a strictly organic and chemical-intensive approach in terms of yield, economics, and environmental benefits. Still, Lotter's not sure we'll get there tomorrow, since the world's farming is hardly pointed in the organic direction—which could be the real problem for the world's poor and hungry. "There is such a huge area in sub-Saharan Africa and South America where the Green Revolution has never made an impact, and it's unlikely that it will for next generation of poor farmers," argues Niels Halberg, the Danish scientist who lead the IFPRI study. "It seems that agro-ecological measures for some of these areas have a beneficial impact on yields and food insecurity. So why not seriously try it out?"

EVOLUTION OF AN ECOLABEL

DAN KENT

The bumper sticker on the old farm truck in the driveway read "Stumps don't lie" as we drove up to the winery in Oregon's Willamette Valley. It was a fall morning and the grape vines looked tired; the fruit had been picked. These were the waning days of the spotted owl–timber wars. Environmentalists and farmers didn't get along much better than environmentalists and loggers. But I saw the bumper sticker and I knew we were on friendly ground. It was 1995 and we were seeking a first farm to test the idea of a new ecolabel called Salmon-Safe which would certify land management practices that protect water quality and wildlife habitat.

Fields and farming were familiar to me. I grew up on a small farm in the rolling fields of eastern Washington's Palouse Hills. Surrounded by wheat farms, I didn't hear much about protecting soils, water, or wildlife. On spring mornings, waiting for the school bus, my sisters and I would feel the cool mist of chemical spray as a crop duster made a pass over the adjacent field.

On my parent's small farm, food was grown in response to the people and farm animals it sustained. As a ten-year-old, I sold potatoes, still covered in dark volcanic soil, from the sliding door of a VW bus. Later in the fall, I'd sell pumpkins, parked on a downtown street in Moscow, Idaho. Following the seasons, I began selling vegetables in a farmers' market: rhubarb in the spring, beans in the summer, tomatoes in the fall. This was the 1970s and I'd sell anything that I could glean from the gardens or orchard.

Once I left the farm, it would be more than a decade before I would think again about agriculture and how food is grown or the possibility that the fate of nature might be intertwined with how we farm. My parents had planted hillsides of Ponderosa pine, elderberry, and dogwood with the same passion that they tended their orchard and livestock. It wasn't until returning home that I saw the habitat they had created for owls, coyotes, and other wildlife through their own relationship to the land.

LAY OF THE LAND

The challenge in starting up the Salmon-Safe certification program wasn't convincing people that it was a good idea. Just about everyone that I talked to along with my colleague, a former USDA Natural Resource Conservation Service agronomist, liked the idea of finding marketplace incentives to reward farmers that protected salmon habitat. The problem was that Salmon-Safe was an idea hatched by an environmental policy think tank called Pacific Rivers Council: a Eugene, Oregon-based organization with a reputation for advocacy and litigation. When we'd talk to farmers they would inevitably express the same hesitation that we often heard from the lawyers and policy experts back in the Pacific Rivers Council office. Somehow we needed to forge a relationship between these urban and rural perspectives: an understanding of the role of farmland in protecting the lowland habitat and its importance to the survival of wild Pacific salmon.

By the 1990s, the plight of wild salmon had become the most strident indicator of declining health in Northwest watersheds. Wild salmon populations had decreased by more than 90 percent in the Columbia River system and native fish had disappeared from 40 percent of their historic Northwest range. Even as a native of the region, I was in my thirties before I witnessed the miracle of wild salmon spawning in the shadow of Oregon's Cascade Range.

That morning in the vineyard marked the real beginning of Salmon-Safe. In the following months, my colleague and I traveled the back roads of the Willamette Valley and elsewhere on the West Coast where we found progressive landowners willing to consider certification under an ecolable that was still very much in development. We ranged as far south as California's Feather River and as far north as Washington's Methow Valley. Mostly we traveled the back roads of the southern Willamette Valley, up against the low gray foothills of Oregon's Coast Range.

Many times, we'd sit with a farmer at the kitchen table, talking about our vision for the ecolable, listening to a description of the difficulties of sustaining a farm in a commodity agricultural system that left farmers powerless in the marketplace. Other times we'd meet with marketing managers or field managers of large organic operations. For these growers, pioneers in the organic movement like Lundberg Family Farms, near Sacramento, or Organic Valley Dairy Cooperative, with several family-run dairies in a lush valley below Washington's Mt. Adams, Salmon-Safe would seem like a natural extension of both their farming and marketing programs. We found that our lineage with the Pacific Rivers Council helped establish credibility with the organic community. Organic farmers became the core audience as we began recruiting growers for certification.

A bigger challenge was finding ways to work meaningfully with commodity-based agriculture. If we really wanted to transform land management practices at the scale required to benefit water quality in salmon watersheds, we had to engage the big dairies, grass seed operators, and wheat farms, not just the low-hanging fruit—small organic farms and boutique wineries. From my experience growing up in wheat country, I knew that this wouldn't be easy. Unlike a bottle of wine, where the name of the farm is typically right on the label, the output from an individual farm growing a commodity crop like wheat or milk is nearly impossible to identify on a grocery store shelf. Our early advances to large industry groups weren't too hopeful. We'd sit on hard plastic chairs in the conference room of a suburban strip mall and try to explain to a politely interested middle manager of a farm cooperative or dairy group why Salmon-Safe certification would benefit their growers. We left one of those meetings, this time with a large dairy cooperative in a freeway office park near Portland, carrying buttons that said "save the spotted cow."

As we worked with uneven success to build landowner momentum, we had yet to actually complete the certification framework that would be essential to our credibility. The name "Salmon-Safe," we were concerned, implied a higher level of protection of salmon than, say, "fish friendly" or "stream safe." Still, we had the sense that our marketplace presence would be amplified by leveraging this totemic icon of the Northwest.

CONNECTING WITH CONSUMERS

Unlike ecolables started by industry groups and foundations, Salmon-Safe has always had a goal of informing the public about the role of farming in the decline of the Northwest's salmon watersheds. From the very start, we communicated this message positively in the now-familiar "vote with your dollars" manner, by promoting products sourced from ecologically sustainable farms. Early in the project, we conducted informal focus groups about the Salmon-Safe logo. We'd place the label on a bottle of Oregon pinot gris and ask consumers what they thought. A more than typical response was that "it must mean that this wine goes well with salmon."

I had no illusions that building a Salmon-Safe brand and generating consumer response to certified products would be swimming downstream. I'd recently completed a marketing MBA at the University of Oregon after a 5-year corporate marketing stint with Bank of America in San Francisco. I'd seen firsthand the cost of mounting ad campaigns that influenced consumer behavior, and I knew we needed a different strategy. Since advertising was out of the question, we decided to invest our scarce financial resources into retail point-of-sale marketing. Now, after years of promoting the logo in supermarkets, I'm gratified when I meet a stranger who says, "Oh yeah, Salmon-Safe. I've seen that logo on a bottle of wine. It means that the vineyard practices reduce runoff into rivers."

This is due in large part to the longtime dedication of a Portland-based ad agency that's worked pro bono on the project from its inception. In the summer of 1995, when Salmon-Safe was still searching for willing farmers and developing the means and measures to certify them, I cold-called Livengood/Nowack, an ad agency that I'd heard might take on a public service project. Their office was up a long flight of stairs. Glancing at an office wall covered with urban graffiti-as-art, I wondered if these were dirt-kicking folks that could talk with farmers. Almost a dozen years later, that same creative team is still developing retail campaigns, print ads, and Salmon-Safe's other marketing and design projects. It's been a two-way street. Based on their Salmon-Safe work, a few years back the agency was awarded the account of one of the nation's largest independent organic food companies.

Ecolabels hadn't yet emerged on the landscape at the time we started Salmon-Safe. Fortunately, there was intense interest from private foundations and government agencies in finding nonregulatory approaches to meeting environmental mandates, particularly on agricultural lands. Organic certification had been around for more than twenty years, and it was just beginning the market ascendance that would make it the $20 billion market it is today.

The only biodiversity-driven ecolable in the marketplace then was Dolphin Safe, established by Earth Island Institute. I traveled to Marin County, just north of San Francisco to talk to the founders of Dolphin Safe. By then they were already working on a Turtle Safe label. But they warned of funding challenges and the difficulties of reaching consumers in an indifferent retail marketplace. I left their office thinking that perhaps the farmlands of the Pacific Northwest were more fertile ground for market-based conservation than the high seas.

With a dozen farm operations prepared to seek certification and our foundation backers calling us to task to get the program into the marketplace, finishing the certification guidelines became our critical task. With support from Pacific Rivers Council's staff salmon biologist and my agronomist colleague on the project, we contracted with an international expert in forestry certification, Robert Hrubes, PhD, to translate the volumes of information we'd gathered about management practices and the ecological needs of salmon into a meaningful certification framework. That first certification framework, relying on a seven-point scale to evaluate the environmental performance of a farm, would be refined endlessly over the next decade. More than 120 farms later, those same standards are still in use today.

The intent of the certification program is to ensure that farmers utilize best management practices to avoid harm, and where appropriate, enhance and restore the health of stream ecosystems. Like organic certification, a Salmon-Safe assessment is based on a comprehensive field inspection with an independent certifier. Unlike organic certification with its singular focus on pesticide inputs, however, Salmon-Safe certification is organized around broader aspects of habitat management. These directly relate to water quality and aquatic habitat protection, riparian area management and erosion control, as well as pesticide reduction and eliminating the use of chemicals that are harmful to aquatic life. But a grower doesn't have to be organically certified to qualify.

INTO THE MARKETPLACE

Salmon-Safe officially launched in March 1997 with a media event at a winery near Portland and the start of a retail campaign in 40 natural food stores, mostly in western Oregon. We had only certified about a dozen products that ranged from packaged rice to organic milk to wine and fresh juice. Amid the contentious stories coming from the Northwest that year of spotted owls and endangered salmon, Salmon-Safe gained attention as a feel-good story. *Newsweek*, *US Today*, and other national media covered the press launch. *Outside* magazine satirized the program with an illustrated container of Salmon-Safe "cookies and stream" ice cream, wondering if "spotted-owl-safe sushi could be far behind." As they say, there's no such thing as bad press.

Even as the label started appearing in Pacific Northwest natural food stores, we began working to gain the confidence of Fred Meyer, then one of the nation's largest independent grocery retailers, to achieve a much larger promotion. A store merchandiser in southern Oregon called to say that people were walking off with our fish-shaped "leaping salmon" shelf talker. "That's a good thing," she then qualified. After six months, and a series of meetings, the company announced that it would promote Salmon-Safe in more than one hundred supermarkets throughout the western United States.

The Fred Meyer launch was the high-water mark of our early years. Oregon's Governor John Kitzhaber, river conservation enthusiast and former Pacific Rivers Council board member, spoke at a media event at a Portland store to kick off the year-long promotion. "Salmon-Safe is the best example yet of the voluntary cooperative approach that is needed to prevent the extinction of wild salmon," said the governor, clad in his trademark Levis and cowboy boots against a backdrop of Salmon-Safe posters and wine bottles.

Salmon-Safe's startup had been funded by two large national foundations, Northwest Area Foundation and Ford Foundation. The foundations had provided Pacific Rivers Council with nearly $500,000 to develop market-based incentives promoting best management practices on agricultural lands. This level of foundation support, we were to learn the hard way, would not be sustainable. By the time the governor spoke on our behalf, Salmon-Safe would be out of funding and surviving on rapidly diminishing subsidies from Pacific Rivers Council.

Meanwhile, I was still following the trail of ecologically sustainable farming as far afield as Idaho's Clearwater River Basin and Washington's Skagit Valley. I'd drive home weary with the realization that change can't happen this slowly across so vast a landscape and still make a difference. We also couldn't charge enough for certification to operate the program without significant foundation subsidies.

I also soon learned that Salmon-Safe's experience was the rule, not the exception. The economics of certification in other regions of the country wasn't penciling out either. Even for nonprofits, farm certification was a break-even business, at best. Organic certifiers, with the 40 percent price premium the organic label then commanded in the supermarket, reported that they were struggling to sustain their operations. One university researcher noted that a large-scale, two-year supermarket campaign for an integrated pest management (known in the industry as IPM) label on apples had resulted in consumers asking, "I PM? Why is the time of picking on the fruit?" If the audience for farm certification had limited potential to fund the important ecological benefits delivered by certification, perhaps we simply needed to find a new audience beyond agriculture.

That opportunity emerged in the form of an inquiry from the city of Portland. In 1999 Chinook salmon in Oregon's Willamette River had been listed as threatened under the Endangered Species Act (ESA). The city of Portland, grappling with the first urban ESA listing in U.S. history, was seeking innovative means to inspire urban citizens, as well as city employees, to reduce water quality impacts in the Willamette River. Their inquiry inspired the development of peer-reviewed certification programs for municipalities and, most recently, corporate and university campuses.

But as Salmon-Safe began to shift toward an exciting expansion into urban watersheds, the project had entirely run out of funding. Foundation supporters of market-based conservation were put off by our unconventional affiliation with a public lands advocacy organization, while traditional environmental funders were not inclined to back a conservation initiative on private land. Our connection with Pacific Rivers Council had also become a barrier to working with municipalities and industry associations; the time had come to separate.

THE NEED TO ADAPT

With $50,000 in seed funding from Pacific Rivers Council and a single employee, Salmon-Safe spun off as an independent organization. At a meeting in Portland's wood-paneled Arlington Club, our founding board charted a new course. We would operate in key salmon watersheds through a network of partnerships with locally based conservation organizations, and we would seek to quantify the impact of the adoption of Salmon-Safe practices in those watersheds.

In southern Oregon, we partnered with Applegate River Watershed Council and World Wildlife Fund to develop a Salmon-Safe Applegate label for family-scale growers of vegetables, livestock, goat cheese, and other agricultural products sourced from this ecologically important tributary to the Rogue River. This template of a locally managed, watershed-based project was next applied in the Snoqualmie Valley, just east of Seattle, in partnership with a Washington-based conservation organization called Stewardship Partners. The Rogue Valley and the Puget Sound region now boast more than 45 Salmon-Safe certified growers, mostly organic farms supplying their local markets.

Building on our long-time work with organic farmers, a next obvious synergy was a partnership with Oregon Tilth, one of the nation's leading certifiers of organic agriculture. While organic growers tend to be among the best farmers in their watersheds, Salmon-Safe has worked with a number of certified operations over the years that haven't met our habitat protection guidelines because of runoff issues, casual use of irrigation water, or inadequate streamside buffers. We collaboratively developed an "overlay" to Oregon Tilth's routine organic inspection that includes additional habitat and biodiversity conservation criteria that are either not covered, or covered only indirectly, under the USDA National Organic Program. Throughout the Pacific Northwest, family-scale organic growers are increasingly seeking to move beyond the national organic program's mandate of eliminating chemical pesticides and fertilizers to differentiate their crops from the rising specter of "industrial organic" in the marketplace. For those growers, the additional Salmon-Safe label conveys that not only is the product healthy, but it's sourced regionally in a manner that protects local watersheds.

In urban centers, Salmon-Safe certifies some of the region's largest corporate and college campuses, including Nike, Toyota, Kettle Foods, and Portland State University. Whether the site is a corporate campus like Nike's with streams and wetlands, an urban office park, or a university campus, certification requires management practices to reduce stormwater runoff and nonpoint source pollution. "Thanks to Nike, they'll run a clean race," says the headline on buses carrying Salmon-Safe public service advertising through downtown Portland in the summer of 2006. The bus side shows salmon swimming with racing bibs.

Still, wine remains Salmon-Safe's flagship product. Working with Oregon Tilth and the Oregon wine industry's sustainability initiative, called Low Input Viticulture & Enology (LIVE), Salmon-Safe has certified 75 vineyards representing a third of Oregon's total vineyard acreage. Early participants, like Brick House, Sokol Blosser, and Bethel Heights, have now been certified for 10 years. Natural food retailers report that sales typically increase 15 to 20 percent during our month-long promotional campaigns, demonstrating the value of the label in the marketplace.

HOW MANY SALMON HAVE WE SAVED?

Quantifying the ecological benefits of practices in the watersheds where Salmon-Safe operates remains a more elusive goal. When reporters call, one of the questions they like to ask is, "How many salmon have you saved?" The facts are grim. Wild salmon populations continue their precipitous decline. Fewer than 5,000 wild Chinook now make the journey up the Snake River drainage of the Palouse Hills farm where I grew up. The Willamette River in Portland, where Salmon-Safe and so many other organizations and individuals have labored for years, still isn't swimable or fishable, much less drinkable.

But in response to the need for quantifiable results, our Applegate program is testing an on-farm ecological monitoring program at five farms in southern Oregon using methods such as vegetation inventories and temperature monitoring. In a few more years, we'll have the data to begin demonstrating the performance of management practices at these sites. In the meantime, walking these farms over the years, I've seen land transformed. Streamside restoration, re-establishment of native vegetation and hedgerows, more efficient irrigation management, livestock fencing, and erosion control measures all establish a more productive farming system where wild nature can thrive.

As I survey the terrain of the Pacific Northwest and northern California, what we now call "salmon nation," I'm hopeful for the future of farming. Though I admit, I also fear that our efforts may be too little, too late for wild salmon. Still, it's important for us to do everything we can to save the salmon, even if, in the back of our minds, we know it's really about more than the salmon. It's about finding meaningful ways to reconnect ourselves to the land, to our watersheds, and to each other. Saving the salmon and saving ourselves will demand nothing less.

WILL AGRICULTURAL ECONOMICS CHANGE IN TIME

DAN BARBER

This Thanksgiving there's something to be really thankful for: More and more Americans are shopping for their turkeys and sweet potatoes at local farmers' markets.

They're doing so because the food is fresher, less processed, and better tasting than what you'd find in a supermarket. But there are also political and social considerations: Supporting small farmers, these shoppers believe, will preserve farmland, reduce the number of industrial farms, and help us move away from an agricultural economy that encourages the production of commodities like corn, soy, and sugar at the expense of just about everything else.

These people are right. And they're also wrong. The bitter truth is that American agriculture—its land and its immensely complex distribution system—is no longer in the hands of the small farmer. Small farmers and farmers' markets, as much as we want them to, are simply not in the position right now to save American agriculture.

Giant farms won't either, of course. For the most part, these are the farms that grow a single crop or raise large numbers of animals in close confinement. To sustain their unnatural existence, these mega-farms, whether they're raising crops or animals, require enormous quantities of pesticides, fertilizers, and antibiotics simply to survive. The result? Pollution, erosion, and diseases that spread easily among factory-raised, immune-deficient animals.

Sadly, these farms aren't going away. In a perverse logic that defies nature, a farm needs to get ever larger and more specialized to survive. The number of farms with annual sales of more than $500,000 increased 23 percent from 1997 to 2002. American farm policy, with a dazzling menu of subsidies, will keep us on this path for the foreseeable future.

The answer to this agricultural puzzle lies somewhere in the middle. Actually, it lies exactly in the middle, with the nation's 350,000 midsize farmers. These farmers, who are too big to sell directly to green markets but too small to compete with highly subsidized industrial farms, cultivate more than 40 percent of our farmland.

Such farmers tend to be highly effective stewards of the land, with intimate knowledge of their farms and their communities. They are small-business owners, not corporations, and have proven records of being interested in protecting not just the economic health of the land, but its ecological health as well.

Unfortunately, these farmers are also on the way out. Midsize farms, with sales of $50,000 to $500,000, are declining rapidly. According to government figures, the number of these farms declined 14 percent from 1997 to 2002, a net loss of nearly 65,000 farms.

According to Fred Kirschenmann of the Leopold Center for Sustainable Agriculture at Iowa State University, it is no longer hard to imagine that most of the farms of the middle will be gone in another decade.

Why should we care? Because our ways of farming are intimately linked to the destructive ways we're eating. Think about your local supermarket. There's fresh produce on the perimeter; but venture into the middle aisles and you're surrounded by processed, canned, preserved, and frozen foods.

It may appear to be a world of variety, but look closer. The cookies, granola bars, crackers, chips, and baby food all have one thing in common: They are made from derivations of corn, soy, and sugar. About 70 percent of our agricultural land in the Midwest is devoted to producing these crops.

The farms that produce these single commodities average about 14,000 acres, roughly the size of Manhattan. And the future? Thomas Dorr, undersecretary of agriculture for rural affairs, has predicted that 250,000-acre behemoths will dominate agriculture. If they do, the number of farms in Dorr's home state, Iowa, would drop to about 120 from 89,000.

That shouldn't come as a surprise. "Get big or get out" has been what farmers have been told for decades. And big farms have come with one big benefit: inexpensive food. Americans spend a smaller percentage of their disposable income for food than anyone else in the developed world. But these savings are illusory.

A funny thing happened on the way to our cheap food system. The books were being cooked in a kind of shell game, Enron-style. The real cost of these monocultures was not being properly accounted for: those taxpayer-financed subsidies ($143 billion during the last decade), the unfairness that results when our excess production gets dumped on developing countries that then can't develop their own resources, the environmental effects of pesticide runoff—the list goes on.

Midsize farms have the potential to be profitable without these hidden costs. After all, there's a large, existing market—school systems, hospitals, local grocery chains, food service distributors—for varied, more healthful foods. These institutions, because of their size, cannot shop at the farmers' market. Even if they could, there would never be enough volume or consistency to meet their needs.

Midsize farms can meet those needs. They may be caught up in the commodity game right now—trying to expand, trying to focus on single crops—but that's largely because that's where the incentives are. For many of these farms, racing to keep up will be their downfall.

We need to encourage these farms to do what they do best: grow a variety of crops, raise a variety of animals, resist the temptation to grow too much.

How do we do this? By shifting the money. Our government now subsidizes the commodity production of grain, mostly corn and soybeans. We need to pull farmers out of the commodity trap and help them make the transition to growing the kinds of whole foods—fruits and vegetables—that would benefit us all. This is not another subsidy, and it's not welfare. It's seed money for a new frontier (actually, an old frontier) in agriculture.

Make no mistake: This change will require us to change our ways. We're going to have to support a diet that contains fewer processed, commodity-based foods. We're going to have to pay more for what we eat. We're going to have to contend with those who question whether it's practical to reduce subsidies for large farmers and food producers. And we're going to have to reward farmers for growing the food we want for our children.

These recommendations may seem bold to the point of audacious. But are they really? After all, what could be more audacious—or contrary to the rural heritage we celebrate this week—than great stretches of our landscape covered with 250,000-acre farms?

A TASTE FOR CONSERVATION

JENNIFER BOGO, ANGIE JABINE,
GRETEL SCHUELLER AND DAVID SEIDEMAN

America's palate is shifting, toward foods that are fresh, sustainable, and chemical free. The chefs leading this culinary crusade believe that eating should be good for both the body and the earth.

From the upscale Savoy in New York City to the casual White Dog Café in Philadelphia, chefs across the country are adding a new ingredient to their dishes: sustainability. They are seeking out food that has been grown locally, thus minimizing resources used to deliver them. Their organic choices are reducing chemicals dumped on the earth. And the food they pick is in sync with the seasons in which it grows, helping support family farms and promoting biodiversity. The result is nothing less than fresh, delicious, nutritious meals, and a perception of wholesome foods that has shifted from granola to gourmet.

What's more, their customers are joining them. Organic foods, for instance, have far outgrown their niche market. In the United States, the organic industry has been expanding by more than 20 percent each year since 1990, reaching sales of $8 billion today. More than a third of the U.S. population now buys organic products everywhere from farm stands and natural food stores to Wal-Mart outlets. The majority of people seek out organic food because—free of chemicals and higher in antioxidants—it's healthier for them. But good taste and a concern for the environment are strong motivators, too.

This change of heart and stomach couldn't have come at a better time: Within the United States, "fresh" produce now travels an average of 1,500 to 2,500 miles before reaching its destination—25 percent farther than in 1980—wasting valuable fuel and polluting the air along its path. Agricultural pesticide use is at record highs; according to the U.S. General Accounting Office, it has risen by almost 40 million pounds since 1992, though the total acreage of cropland has decreased. A mere 10 to 15 species of plants and 8 species of livestock now account for 90 percent of the world's food production, and that range is narrowing. This drop in diversity deprives the nation of a genetic arsenal against pests and disease.

By artfully changing the cuisine on their plates, professional chefs have been able to promote a new social and environmental consciousness. As the nation's taste-shapers—turning once-exotic items like mesclun greens into fare as mainstream as iceberg lettuce—chefs often set the example for the food people put on their dinner tables. Every day, the cult of the "celebrity chef" is played out in glossy lifestyle magazines, television food programs, and Internet sites. Americans are eating out at an all-time high—more than 30 percent of meals are now consumed away from home.

Furthermore, the restaurants themselves represent huge purchasing power. "We buy millions of dollars' worth of food each year," says celebrity chef Rick Bayless, winner of several distinguished culinary awards, host of a TV cooking show, and owner of two Mexican-themed restaurants in Chicago. His restaurants alone spend more than $300,000 on produce from local farms. "We are giving them a huge financial shot in the arm to boost organic farming."

Ten years ago Bayless helped found the Chefs Collaborative, which arms its membership of 1,000 restaurants and chefs with the tools for making environmentally sound purchasing decisions. "We found that many of the chefs organizations weren't addressing the issues we were dealing with. Foods were being promoted that aren't sustainable," says Bayless. Unlike many professional culinary organizations—focused more on the innovative presentation of foods—members of the Chefs Collaborative started to examine where their ingredients were coming from, and how to improve upon that record: Did sea turtles get trapped and killed in nets in order to serve that shrimp cocktail? What are good seafood substitutions

Organic: Clean Cuisine

Jennifer Bogo

Somewhere behind the rising columns of steam in Nora Pouillon's kitchen, tempura squash blossoms, oven-roasted sweet peppers, and slow-braised rabbit await finishing touches; shiitake-tofu stuffing, black-olive dressing, and cognac-mustard sauce materialize from the silvery flash of knives and whisks. Anyone first ushered into the softly lit dining room of Restaurant Nora, its walls elegantly hung with antique quilts, may find the epicurean menu unsurprising—if not for one notable detail: Everything, down to the herb aioli, is organic.

"People hear the word organic and think it means they have to be on a special diet, to restrain themselves, to eat beans and rice," grouses Pouillon in her native Austrian accent. "People eat the exact same thing here as they would in any upscale restaurant. To be organic is to be better for you and for the environment."

In 1999 Restaurant Nora, in Washington, D.C., became the first restaurant in the nation to be certified organic, which means at least 95 percent of the ingredients it uses must be certified organic, too. Organic farmers—whether they're raising ducks or daikon—cannot add chemical pesticides or fertilizers. Nor can they produce foods that are genetically engineered or grown using hormones or sewage sludge. They can and do, however, build the long-term health of their soil, protect water quality, and foster ecosystems friendly to wildlife.

This is revolutionary in a country that uses more than 500 million pounds of pesticides and 21 million tons of chemical fertilizers to produce its food, and pollutes more than 173,000 miles of waterways in the process of growing it. And yet long-term studies have shown that organic crops can actually outproduce those conventionally grown, particularly during conditions like drought. Organic farming also requires 50 percent less energy than conventional methods, reducing greenhouse-gas emissions, and can help to offset global warming by locking even more carbon into the soil.

Organic farming, which substitutes labor-intensive management for the quick fix of chemicals, does cost more, but the price tag better reflects the true expense of growing our nation's food. "People are disconnected with where their food comes from," says Pouillon, her hands accentuating each point. "It doesn't grow in the supermarket. There's a person called a farmer that works really hard to grow the grain that is perhaps your pasta, or who milks the cows twice every day to give you butter."

Pouillon's passion has in no way slowed her success: She has been named Chef of the Year by both the International Association of Culinary Professionals and the American Tasting Institute. And in the nation's capital, where even minor change requires a lobbyist and an act of Congress, Restaurant Nora and its "new" organic cooking is making waves. Environmentalists, tourists, celebrities, politicians, CEOs, and, yes, editors add their names to the reservations list, hoping to sink their spoons into Nora's warm chocolate soufflé cake. The moment that rich, organic chocolate melts into a pool of homemade ice cream, they're sold. Pouillon can't help but laugh: "People always thought I was nuts, but now they have caught up."

for overfished species? The organization, he says, is "answering questions that many of us had been asking for a long time."

Furnished with this information, these restaurateurs are taking their role as educators as seriously as the food they serve. Take, for example, the Putney Inn, a historic farmhouse turned restaurant in southern Vermont, where 100,000 people each year—many on their way to view the fall foliage—stop to savor a meal. Even in this quaint, out-of-the-way restaurant, a small revolution is cooking.

On the menu's first page, the inn outlines its commitment to offer the "least processed products possible" and to support local farms. The ingredients have actual origins: native, organic lamb from a guy named Frank, honey made right in Putney. Most of the food comes from less than three hours away, and most of the time about 85 percent of the produce is organic, says Kevin Takei, the inn's executive chef and a member of the Vermont Fresh Network, which connects chefs with local farms. "Practically everything I use is from New England." He often shops from the local farm stands to buy the latest pickings. "The farmers literally will harvest it that day: spinach, beets, squash, sweet corn. You see what's avail-

Meat: High Steaks

Angie Jabine

If you can't finish the meltingly tender flatiron steak at Higgins Restaurant in Portland, Oregon, you shouldn't hesitate to ask for the leftovers. Nationally acclaimed chef Greg Higgins wants you to enjoy every bite. He has gone out of his way to ensure that all of the foods in his kitchen—even the meat—come from local, sustainable farms. "To me these terms have become synonymous with quality," says Higgins. "I want my suppliers to be passionate people who love what they do, because I find their products are head and shoulders above conventional."

Take that flatiron steak. It came from Oregon Country Beef, a cooperative of 40 ranchers who manage their eastern Oregon rangeland in ecologically responsible ways and raise their cattle without growth hormones, antibiotics, or bioengineered feed. Higgins also carries as much grass-fed, certified-organic beef as he can from Jim and Ellen Girt of River Run Farm in nearby Clatskanie. The restaurant's pork-shoulder terrine—a robust yet mild preparation enlivened with tart, dried Northwest cherries—is made from pork that's been pasture-raised by Sara and Joe DeLong, organic farmers in St. John, Washington.

Higgins is in the vanguard of a steadily growing movement to support local producers who want to protect the long-term health of the planet. By the time the average feedlot steer is ready for slaughter, it has eaten 1,900 pounds of grain. Its herd has generated enormous amounts of waste, which, collected and spread over nearby fields, leaches excess nitrogen into the soil, streams, and groundwater. Pasture-raised livestock and poultry, by contrast, fertilize the land as they graze it, and don't require fuel-intensive tractors to harvest grain for year-round feed.

Though not all sustainably produced meat necessarily carries the certified-organic label, that which does guarantees antibiotics haven't been added to the animals' feed—a practice that has contributed to antibiotic resistance in people. The label also means that the livestock or poultry has been raised on feed that is organic, so both pastures and fields used to grow supplemental grain have been cultivated without chemical fertilizers or pesticides.

What's more, unlike with the Washington dairy cow found infected with mad cow disease last winter, the herd and feed history of an organically raised steer is quickly traceable. The animal has also been slaughtered and processed separately from conventional livestock to avoid cross-contamination.

As a restaurateur–and as a realist–Higgins concedes there is not enough grass to sustain a meat supply at the rate Americans consume it. And it's true that certified-organic, grass-fed products are more expensive–partly because of higher record-keeping costs and partly because of the economies of scale favoring big business. But when you consider the environmental benefits, and the fact that these animals are free of the chemicals that conventionally raised animals store in their bodies, the price seems well worth it.

able, then you create the menu," Takei explains. Toward the end of the evening, owner Randi Zeitler and Takei will often make rounds to the tables and talk about the food and where it comes from. "With family farms at an all-time low and agribusinesses increasing, teaching customers about who grew their food is important," says Takei.

Of course, there's still what Bayless calls the "lazy chef," who calls up a supplier in winter and asks for a flat of strawberries, without a thought to how many pesticides it's taken to grow them or how far they'll have to travel. But the tide is turning. "We're just one link in the food system, from the people who till the soil to the people who eat the food," says Bayless. But as the profiles of chefs Nora Pouillon, Greg Higgins, Annie Somerville, and Rick Moonen illustrate, they are a very powerful link. Their choices, like those of the customers they influence, can help preserve and promote an ecologically sound system of agriculture. "We're dedicated to making our country's food better," Bayless says, "for us and our environment."

Vegetarian: Field of Greens

Jennifer Bogo

Annie Somerville and I weave between rows of lettuce—curly endive, butter, red romaine—their plump heads crowding one another under the California sun. "Aren't these great?" she exclaims, leaning down to rub a furrowed pink leaf between her fingers. "This is a variety of red oak. As it gets bigger, it gets darker and more intense looking. Vinaigrette gets caught in the crinkles."

We are tasting our way around Green Gulch Farm, tearing off bits of aniselike fennel, pungent lemony chard, and peppery arugula as we walk. The farm, nestled along the coast just 17 miles north of San Francisco, provides year-round organic produce to the city's highly praised Greens Restaurant, where Somerville is executive chef.

Diners who think vegetarian restaurants have little more to offer than tofu and alfalfa sprouts have only to take in the confettilike stems of rainbow chard, the long, thin, red blossoms of pineapple sage, and the lavender, silver and fuzzy, that fill the fields around us. "We just cook in the spirit of using really great ingredients," says Somerville. "And since we are a vegetarian restaurant, we're not thinking about what fisheries are being overfished or what the conditions are for livestock. In terms of resource use, our restaurant is really very light."

There's no question that raising meat, fish, and fowl is far more wasteful and energy-intensive than growing food for a vegetarian diet. Compared with a pound of pasta, for instance, a pound of red meat is responsible for 20 times the land use, 17 times the water pollution, 5 times the water use, and 3 times the greenhouse-gas emissions. Animals' bodies also concentrate growth hormones, antibiotics, and pesticides, passing them on to ours when we eat them. Many Greens customers are vegetarians, says Somerville, but many are also omnivores who come to the restaurant simply for a really good meal. If everyone followed suit, skipping just one meat-based meal a week, the amount of resources saved would be staggering.

Through Greens, Somerville shows just how easy, and delicious, that choice can be. Back in the restaurant's airy dining room, in the Fort Mason Complex in the Golden Gate National Recreation Area, a wall of windows provides a panoramic view of the Marin Headlands and the Golden Gate Bridge. Late-afternoon light illuminates Somerville's face—well tanned from long days spent outdoors—as she ponders aloud her favorite food.

"I *love* Thai basil, porcini mushrooms, all the potatoes and onions," she says, with evident zeal. "The tender shoots of green garlic are great on pizza and pasta. And Meyer lemons are sweet, like a regular lemon that's been crossed with an orange." She leans back, ensconced now in a kitchen far away. "Poblano chilies have a smoky, wonderful deep flavor. They're good with plantains, which are also great in winter stews. And quince is a really aromatic old-world fruit."

People fill the tables around us as we talk. My gratin soon arrives, a tower of perfectly grilled summer squash topped by a golden crust, the heavenly aroma of sun-soaked vegetables wafting skyward. It is a perfect testament to the indulgence of eating food reaped straight from the earth.

Seafood: Fish on the Line

David Seideman

"His energetic style causes the masses to crave his seafood specialties," gushes the website Starchefs.com. "We call him King of the Sea." Conservationists may as well dub Rick Moonen savior of the seas. "I truly believe that to be part of eradicating a species from the water or anywhere is just plain irresponsible," he says. "And nothing I want to be part of."

Moonen, owner of Restaurant rm on Manhattan's Upper East Side, is a fixture on popular television shows such as *Today* and *Good Morning America,* as well as on the Food Network. New York City's toughest food critics shower him with stars. At the same time he has exploited his celebrity status to preach what he practices.

Moonen proclaims he serves sustainable seafood not just to maintain his future supplies but because it's good for the earth. After all, the U.S. government warns that almost 100 fish populations are overtaxed. The journal *Nature* reports the earth has lost 90 percent of its large predatory fish species, such as cod, flounder, and tuna. Meanwhile, the U.S. Department of Agriculture estimates that fish consumption will jump 7 percent by 2020.

Restaurant rm evokes a sleek mahogany yacht, with portholes on the kitchen doors and a raftered ceiling that looks like the ribs of a ship. For an appetizer, fresh East Coast oysters melt in the mouth. "Oysters are a great choice," Moonen says, a smile flashing across his intense gaze. "They clean the ocean." Indeed, oysters filter water and can be farmed at negligible cost to the environment. For an entrée, the seared Arctic char, topped with a crisp skin and served with roasted organic beets and red-wine–beet vinaigrette, is succulent. Moonen resists advertising the species' healthy population: "The concern is that if you send out a message that Arctic char is a viable alternative to farm-raised Atlantic salmon, guess what? In a very short time it would be gone."

Moonen's Frisée au Salmon has a tender but chewy texture and offers a slight kick from being cured with bacon. The salmon is wild; the farm-raised variety causes pollution and lacks genetic diversity. Moreover, a recent study in *Science* found that it has an unhealthy concentration of toxins.

There are many other popular species Moonen won't serve, including Chilean sea bass, orange roughy, and shrimp (except as a garnish on pasta). These species are overfished, they take too long to mature, they struggle from degraded habitat, or they're caught in a way that leads to the incidental kill of a variety of marine life.

But there is another species, long absent from Moonen's menu, that he could put back—if he still had a taste for it. Six years ago Moonen emerged as chief spokesman for the Give Swordfish a Break campaign, led by chefs and conservationists. Thanks to a public boycott and fishing regulations, swordfish has made a tenuous recovery. Being able to serve it again "gives some sort of completion to an effort," Moonen says. "You can say, 'You showed your concern...and here it is back for you to enjoy.' If everything is met with dismal finality, then people just get tired of it, and they stop getting involved."

CONTRIBUTORS

Dan Barber is the executive chef and co-owner of Blue Hill at Stone Barns as well as the creative director of the Stone Barns Center for Food and Agriculture in Pocantico, New York. An acclaimed chef who focuses on regionally produced foods and the importance of everyday food choices, Barber also serves on Harvard Medical School's Center for Health and the Global Environment advisory board and works on policy issues with the Kellogg Foundation, New York City's Green Markets, and Slow Food USA.

Rick Bass is the author of 22 books of fiction and nonfiction, including, most recently, a short story collection, *The Lives of Rocks.* He grew up in Texas, studied wildlife science and geology at Utah State University, and worked as a biologist in Arkansas and a geologist in Mississippi and Alabama. He lives in northwest Montana's Yaak Valley, where he serves as a board member of the Yaak Valley Forest Council (www.yaakvalley.org), working to help protect as wilderness the last unprotected roadless areas in the Yaak Valley—the Land the Wilderness Act Forgot.

Jo Ann Baumgartner is director of the Wild Farm Alliance, whose mission is to promote a healthy, viable agriculture that protects and restores wild Nature. She was an organic farmer for over a decade, has studied bird predation of insects in orchards, worked in organic cotton, and researched endangered species in California.

Wendell Berry is a working farmer in north central Kentucky and the author of more than 30 books of poetry, essays, and novels. He serves on the Wild Farm Alliance advisory board and has received fellowships from the Guggenheim and Rockefeller Foundations, a Lannan Foundation Award, and a grant from the National Endowment for the Arts.

Jennifer Bogo is the senior editor for science at *Popular Mechanics* magazine, where she writes and edits stories on topics ranging from nuclear power to the exploration of Saturn. She previously was an editor at *Audubon* magazine where she assigned the story "A Taste for Conservation" included in this reader and at *E/The Environmental Magazine*. She was trained in environmental science and biology.

John Davis serves on the boards of the Wildlands Project, RESTORE: The North Woods, Wild Farm Alliance, and Eddy Foundation, and as a fellow of the Rewilding Institute, and as conservation director of the Adirondack Council. In the Adirondacks, where he and his family reside, he and colleagues are piecing back together a wildlife corridor—which will be buffered by small, wildlife-friendly farms—linking the Champlain Valley with the High Peaks.

Dave Foreman is director of the Rewilding Institute and author of *Rewilding North America.* He was a cofounder of the Wildlands Project and Earth First! He worked for the Wilderness Society in the 1970s and is currently a board member of the New Mexico Wilderness Alliance and an advisory board member of the Wild Farm Alliance. *Audubon* magazine honored him as one of its 100 Champions of Conservation in the 20th Century.

David Gould has a background in Life Sciences from MIT and 10 years' work in alternative education. For the past 11 years he has helped to form standards and policy in the production and certification of organic, Fair Trade, non-GMO, and local food systems. He has conducted thousands of inspections and certifications, and trained regulatory professionals, graduate students, government officials, and producers on six continents. David is currently pursuing his own permaculture dream on 10 acres in coastal Oregon.

Brian Halweil is a Senior Researcher with the Worldwatch Institute where he writes on the social and ecological impacts of how we grow food, focusing recently on organic farming, biotechnology, hunger, and rural communities. He describes the evolving local food movement in *Eat Here: Reclaiming Homegrown Pleasures in a Global Supermarket.* He writes from Sag Harbor, NY, where he and his wife tend a home garden and orchard.

Daniel Imhoff is the author and publisher of numerous books, including *Farming with the Wild, Paper or Plastic, Building with Vision,* and more recently, *Food Fight: The Citizen's Guide to the Food and Farm Bill.* In addition, he is a sought-after speaker and lecturer, cohosts a sustainable agriculture radio show, runs a biodynamically certified small-scale farm, and is a cofounder and current president of the Wild Farm Alliance.

Angie Jabine spent three years as a field archaeologist before becoming a journalist. She has contributed stories and reviews to *Willamette Week*, the *Oregonian*, the *New York Times*, the *Washington Post*, *Advertising Age*, *Audubon*, and other publications. She is editor of *Northwest Palate*, the magazine of food, wine, and travel in the Pacific Northwest. She lives in Portland, Oregon, with her husband and their two daughters.

Laura L. Jackson is professor of biology at the University of Northern Iowa in Cedar Falls where she teaches courses in ecology, conservation biology, and environmental studies. She received a bachelor's degree in biology from Grinnell College, and a PhD in ecology from Cornell University. She is the coeditor with Dana Jackson of *The Farm as Natural Habitat: Reconnecting Food Systems and Ecosystems*, published by Island Press (2002).

Dan Kent directs Salmon-Safe, an organization devoted to restoring water quality and salmon habitat in agricultural and urban landscapes of the Pacific Northwest. He serves on the board of directors of the Wild Farm Alliance.

Barbara Kingsolver's 11 books include essay collections, short stories, poetry, an oral history, and many well-known novels including *The Bean Trees* and *The Poisonwood Bible*. She and her husband, Steven Hopp, and their two daughters grow most of their own food on a farm in southern Appalachia. Her newest book, *Animal, Vegetable, Miracle* is forthcoming in May 2007.

Fred Kirschenmann is the Distinguished Fellow of the Leopold Center for Sustainable Agriculture and president of Kirschenmann Family Farms, a 3,500-acre certified organic farm in Windsor, North Dakota. He is a Wild Farm Alliance advisory board member, past-president of Farm Verified Organic, and has served on the U.S. Department of Agriculture's National Organic Standards Board.

Aldo Leopold (January 11, 1887-April 21, 1948) was the author of *A Sand County Almanac* and *Sketches Here and There* and was a United States forester and environmentalist. To many he is known as the father of modern ecology and was influential in the development of modern environmental ethics and conservation. Leopold led a generation to a new perception of nature and to a new vision of relationship with the natural environment.

Richard Manning is the author of seven books including *Against the Grain,* which makes the case that 10,000 years of agriculture have damaged nature and human nature; *Food's Frontier,* a profile of nine post–Green-Revolution agricultural projects in Latin America, Asia, and Africa; and *Inside Passage,* an account of biodiversity and economy in the coastal temperate rainforests of North America. He is also a freelance magazine writer, with essays and articles published in *Harper's, Wired,* the *Los Angeles Times, American Scholar, Frankfurter Allgemeine Zeitung,* the *New York Times, Audubon,* among others.

Scott McMillion has covered the environment and natural resources for the *Bozeman Daily Chronicle* for 18 years. He is the author of the award-winning book *Mark of the Grizzly* and lives in Livingston, Montana.

Gary Paul Nabhan, Director of the Center for Sustainable Environments at Northern Arizona University, is cofounder of Native Seeds/SEARCH and a Wild Farm Alliance advisory board member, has served as director of conservation science at the Arizona-Sonora Desert Museum in Tucson, and has received a MacArthur "Genius" Fellowship Grant. He crosses disciplinary, cultural, and ethnic boundaries to work with many different communities in the Southwest.

Reed Noss is the Davis-Shine Professor of Conservation Biology at the University of Central Florida and the chief scientist for the Wildlands Project. He has authored numerous articles and several books related to conservation biology and is a leading expert on large-scale conservation strategies. He serves on the Wild Farm Alliance advisory board.

Michael Pollan is the author of *The Omnivore's Dilemma, Second Nature, A Place of My Own,* and *The Botany of Desire,* a *New York Times* bestseller. He is a longtime contributing writer at the *New York Times* magazine and teaches journalism at the University of California-Berkeley.

Gretel Scheuller writes about science and the environment. Her articles have appeared in *Audubon* magazine, *Discover, Hooked on the Outdoors, National Wildlife, New Scientist, Popular Science, SKI,* and on ENN.com. Before becoming a freelance writer, she held editorial positions at several national magazines, including *Audubon* and *Earth*. Currently, she is working on a book about the environmental and cultural effects of food. She also teaches journalism at the State University of New York in Plattsburgh.

David Seideman is editor-in-chief of *Audubon* magazine, where he strives to broaden the base of their support by covering unexpected allies in the movement, whether deer hunters, evangelical Christians, farmers, or land developers. He has also worked at *The New Republic* and *Time* magazines.

John Terborgh is a professor at the Nicholas School of the Environment at Duke University and has a BA, MA, and PhD from Harvard University. His interests lie in the fields of tropical ecology and conservation. The common denominator in all of his work has been the goal of solving problems of general ecological interest using a comparative approach.

Ana Guadalupe Valenzuela-Zapata is recognized in Mexico and the United States as the foremost botanical and horticultural expert on agaves used in the tequila industry. A professor at the Universidad de Guadalajara and one of the few women field scientists in a male-dominated industry, she is author of the Spanish-language book *El Agave Tequilero.*

Luba Vangelova is a freelance journalist living in Washington, DC. She has written about the environment, travel, social issues, science, and other topics for publications such as *Smithsonian, National Geographic Traveler,* the *New York Times,* and *Wildlife Conservation*. She has also contributed to four books.

Becky Weed is co-owner of certified organic Thirteen Mile Lamb and Wool Company, which manages sheep under the Predator Friendly label without killing coyotes, mountain lions, bears, eagles, and wolves. She is a Wild Farm Alliance board member.

Ted Williams has been a full-time writer on environmental issues, with special attention to fish and wildlife conservation, since 1970. In addition to freelancing for national magazines, he contributes regular feature-length conservation columns to *Audubon* and *Fly Rod & Reel* where he serves as Editor-at-Large and Conservation Editor respectively.

SELECTED RESOURCES

Baskin, Y. *The Work of Nature: How the Diversity of Life Sustains Us.* Washington, DC: Island Press, 1997.

Berman, M. *The Reenchantment of the World.* Ithaca: Cornell University Press, 1981.

Berry, W. *Another Turn of the Crank.* Washington, DC: Counterpoint, 1995.

______. *Citizenship Papers.* Emeryville, CA: Shoemaker & Hoard, 2004.

______. *The Unsettling of America: Culture and Agriculture.* Berkeley: University of California Press, 1977.

Boody, G., B. Vondracek, D. A. Andow, M. Krinke, J. Westra, J. Zimmerman, and P. Welle. Multifunctional Agriculture in the United States. *BioScience* 55:27–38 (2005).

Buchanan, M. *Nexus: Small Worlds and the Groundbreaking Science of Networks.* New York: W. W. Norton & Company, 2002.

Buchmann, S., and G. P. Nabhan. *The Forgotten Pollinators.* Washington, DC: Island Press, 1996.

Carson, R. *Silent Spring.* Boston: Houghton Mifflin, 1962.

Catton, W. *Overshoot.* Champaign, IL: Univ. of Illinois Press, 1982.

Cronon, W. *Changes in the Land: Indians, Colonists and the Ecology of New England.* New York: Hill and Wang, 1983.

Dinnes, D. L., D. L. Karlen, D. B. Jaynes, T. C. Kaspar, J. L. Hatfield, T. S. Colvin, and C. A. Cambardella. Nitrogen Management Strategies to Reduce Nitrate Leaching in Tile-Drained Midwestern Soils. *Agronomy Journal* 94:153–171 (2002).

Ehrenfeld, D. *The Arrogance of Humanism.* New York and London: Oxford Press, 1981.

Eldredge, N. *Dominion: Can Nature and Culture Co-Exist?* New York: Henry Holt and Company, 1995.

Foreman, D. *Rewilding North America: A Vision for Conservation in the 21st Century.* Washington, DC: Island Press, 2004.

Hendrickson, M. K., and H. S. James, Jr. The Ethics of Constrained Choice: How the Industrialization of Agriculture Impacts Farming and Farmer Behavior. *Journal of Agricultural and Environmental Ethics* 18:269–291 (2005).

House, F. *Totem Salmon: Life Lessons from Another Species.* Boston: Beacon Press, 1999.

Howard, Sir A. *An Agriculture Testament.* New York and London: Oxford University Press, 1943.

Imhoff, D. *Farming with the Wild: Enhancing Biodiversity on Farms and Ranches.* San Francisco: Sierra Club Books/Healdsburg, CA: Watershed Media, 2003.

Jackson, D. L., and L. L. Jackson, eds. *The Farm as Natural Habitat: Reconnecting Food Systems with Ecosystems.* Washington, DC: Island Press, 2002.

Kauffman, S. A. *The Origins of Order.* New York and London: Oxford University Press, 1993.

Kingsolver, B. *Small Wonder Essays.* New York: HarperCollins, 2002.

Kunstler, J. H. *Long Emergency: Surviving the Converging Catastrophes of the Twenty-First Century.* New York: Atlantic Monthly Press, 2005.

Leopold, A. "A Biotic View of the Land," in Susan L. Flader and J. Baird Callicott (eds.) 1991. *The River of the Mother of God and Other Essays by Aldo Leopold.* Madison: University of Wisconsin Press, 1939.

______. *A Sand County Almanac.* New York and London: Oxford University Press, 1949.

Logsdon, G. *All Flesh Is Grass: The Pleasures and Promises of Pasture Farming.* Athens, OH: Swallow Press, 2004.

Mander, J., and E. Goldsmith, eds. *The Case Against the Global Economy.* San Francisco: Sierra Club Books, 1996.

Manning, R. *Grassland: The History, Biology, Politics, and Promise of the American Prairie.* New York: Penguin Books, 1995.

Nabhan, G. P. *Coming Home to Eat: The Pleasures and Politics of Local Foods.* New York: W. W. Norton and Company, 2002.

______. *The Desert Smells Like Rain: A Naturalist in O'Odham Country.* Tuscon: The University of Arizona Press, 1982.

Nabhan, G. P., and S. Trimble. *Cultures of Habitat: On Nature, Culture, and Story.* Washington, DC: Counterpoint, 1994.

Nabhan, G. P., and A. Valenzuela-Zapata. *Tequila: A Natural and Cultural History.* Tucson: The University of Arizona Press, 2004.

Noss, R. F. *Saving Nature's Legacy: Protecting and Restoring Biodiversity.* Washington, DC: Island Press, 1994.

Pimm, S. *The World According to Pimm: A Scientist Audits the Earth.* New York: McGraw-Hill, 2001.

Pollan, M. *The Omnivore's Dilemma: A Natural History of Four Meals.* New York: Penguin Press, 2006.

Rasmussen, L. *Earth Community, Earth Ethics.* Maryknoll, NY: Orbis Books, 1996.

Reisner, M. *Cadillac Desert: The American West and Its Disappearing Water.* New York: Penguin Press, 1986.

Robinson, J. *Pasture Perfect: The Far-Reaching Benefits of Choosing Meat, Eggs, and Dairy Products from Grass-Fed Animals.* Vashon, WA: Vashon Island Press, 2004.

Schlosser, E. *Fast Food Nation: The Dark Side of the All-American Meal.* Boston: Houghton Mifflin, 2001.

Shepherd, M., S. L. Buchmann, M. Vaughan, and S. Hoffman Black. *Pollinator Conservation Handbook.* Portland: The Xerces Society, 2003.

Soulé, M., and J. Terborgh, eds. *Continental Conservation.* Washington, DC: Island Press, 1991.

Stein, B. A., L. S. Kutner, and J. S. Adams. *Precious Heritage: The Status of Biodiversity in the United States.* New York and London: Oxford University Press, 2002.

Wilcove, D. S. *The Condor's Shadow: The Loss and Recovery of Wildlife in America.* New York: W. H. Freeman and Co, 1999.

Wilson, E. O. *The Diversity of Life.* Cambridge, MA: Harvard University Press, 1992.

Wolf, E. C., and S. Zuckerman, eds. *Salmon Nation.* Portland: Ecotrust, 2003.

INDEX

F

G

H

M

N

O

R

S

X

Y

Z

ILLUSTRATION CREDITS

The following illustrations have been used with permission from their artists, and all rights are retained by them.

Page 3 Hay bales: Gertrude Ten Broeck

Page 14 Swainson's hawk: Paul Kerris, U. S. Fish and Wildlife Service

Page 38 Ocelot: Bob Savannah, U. S. Fish and Wildlife Service

Page 48 Crop rows: Natural Resources Conservation Service

Page 63 Salmon: Brian Beard, San Anselmo, CA

Page 73 Mexican long-tongued bat: George Malesky, Pima County, AZ

Page 80 Beaver: Charles Douglas, Canadian Museum of Nature, Ottawa Canada

Page 84 Grass: Washington Department of Fish and Wildlife

Page 91 Wolf: Darrell Pruett, Washington Department of Fish and Wildlife

Page 96 Bee: Charles Douglas, Canadian Museum of Nature, Ottawa Canada

Page 103 Dragonfly: Roger Hall, www.inkart.net

Page 117 Farm scene with tree: JupiterImages Corporation © 2006

Page 126 Bear cubs: Oregon Department of Fish and Wildlife

Page 131 Ovenbird: Roger Hall, www.inkart.net

Page 143 Bear print: Washington Department of Fish and Wildlife

Page 152 Cougar: Bob Savannah, U. S. Fish and Wildlife Service

Page 166 Maple leaves: JupiterImages Corporation © 2006

Page 179 Corn field: William Crook, Jr., Springfield, Il

Page 187 Peas: JupiterImages Corporation © 2006

Page 200 Pond scene: Anne Ziller, Eugene, OR

Page 210 Farm stand: Roland Lee, St. George, UT © 2003

ABOUT THE WILD FARM ALLIANCE

The Wild Farm Alliance promotes a healthy, viable agriculture that helps protect and restore wild Nature. Our vision shifts the focus beyond the isolated, individual farm to one that includes the natural ecosystems of the region. By farming with the wild, agriculture can provide wildlife habitat within its borders, and connections to wildlands beyond—through the creation of buffers and corridors that are permeable to the movements of animals and that help to link people with the land.

The Wild Farm Alliance Briefing Papers Series explores many of the issues that define and distinguish the concept of conservation-based agriculture. Each paper focuses on a particular issue set in the context of reconnecting food systems with ecosystems.

- *Farming with the Wild Forever: Using Agricultural Easements to Support Biodiversity*
- *Making Connections for Nature: The Conservation Value of Farming with the Wild*
- *Water: Life Blood of the Landscape*
- *Local Control in the Global Arena: Restructuring Ecological Food Systems for the Protection of Natural and Human Communities*
- *Linking Conservation with the Bottom Line: Incentives for Farming with Nature*
- *Grazing for Biodiversity: The Co-Existence of Farm Animals and Native Species*
- *Agricultural Cropping Patterns: Integrating Wild Margins*

Our biodiversity conservation guides lay out a range of farm management possibilities for a variety of situations that maintain and enhance biodiversity at the farm level and contribute to biodiversity conservation outside of farm borders at the regional or watershed level.

- *Biodiversity Conservation: An Organic Farmer's Guide*
- *Biodiversity Conservation: An Organic Certifier's Guide*

The Guides and the briefing papers can be downloaded from our website.

For more information about our programs, please contact:

PO Box 2570
Watsonville, CA 95077
(831) 761-8408
www.wildfarmalliance.org

OTHER BOOKS BY WATERSHED MEDIA

Paper or Plastic: Searching for Solutions to an Overpackaged World
Written by Daniel Imhoff
Introduction by Randy Hayes
ISBN-1-57805-117-1, Co-published with Sierra Club Books in 2005
168 pages, 100 original photographs
US$ 16.95

Farming with the Wild: Enhancing Biodiversity on Farms and Ranches
Written by Daniel Imhoff
Designed by Roberto Carra
Introduction by Fred Kirschenmann
ISBN-1-57805-092-8, Co-published with Sierra Club Books in 2003
184 pages, 200 original photographs
US$ 29.95

Building with Vision: Optimizing and Finding Alternatives to Wood
Written by Daniel Imhoff
Designed by Roberto Carra
Introduction by Sim Van der Ryn
ISBN-0-9709500-0-4, Published in 2001
136 pages, 200 original photographs
US$ 22.00

Food Fight: The Citizen's Guide to a Food and Farm Bill
Written by Daniel Imhoff
Foreword by Michael Pollan
Introduction by Fred Kirschenmann
ISBN 0-9709500-2-0, Due in early 2007
US$ 16.95

Resources for the Transition to a Sustainable Society
451 Hudson Street, Healdsburg, California 95448
(707) 431-2936 www.watershedmedia.org
Watershed Media Books are distributed by the University of California Press

WILD FARM ALLIANCE

Board of Directors & Advisors